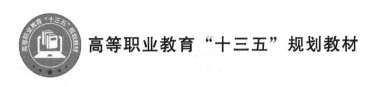

高等职业教育"十三五"规划教材

航空英语

主　编　那　森　井丽莉
副主编　杨　策　白春花

北京邮电大学出版社
www.buptpress.com

内 容 简 介

本书是以典型性的地面服务和客舱服务工作任务为主线，结合地勤工作、空中服务工作岗位的职业规范，以及学生自身的认知特点，以掌握其服务技能的英语沟通要点而设计的一体化教材。本教材以地勤工作人员和空中服务人员在实际工作中遇到的具体情境为任务，展开一系列的学习活动设计。每个学习活动下的工作任务根据主题设计了相应的学习任务和情景，以培养学生的综合职业能力。本教材适合职业院校航空服务类专业及其相关专业学生学习、培训使用。

图书在版编目（CIP）数据

航空英语 / 那淼，井丽莉主编 . -- 北京：北京邮电大学出版社，2020.7（2024.1重印）
ISBN 978-7-5635-6106-3

Ⅰ．①航… Ⅱ．①那…②井… Ⅲ．①航空—英语 Ⅳ．①V2

中国版本图书馆 CIP 数据核字（2020）第 108509 号

策划编辑：张向杰　　　　　　责任编辑：廖 娟　　　　　　封面设计：七星博纳
出版发行：北京邮电大学出版社
社　　　址：北京市海淀区西土城路 10 号
邮政编码：100876
发 行 部：电话：010-62282185　　传真：010-62283578
E-mail：publish@bupt.edu.cn
经　　销：各地新华书店
印　　刷：北京虎彩文化传播有限公司
开　　本：787 mm×1 092 mm　1/16
印　　张：7
字　　数：150 千字
版　　次：2020 年 7 月第 1 版
印　　次：2024 年 1 月第 3 次印刷

ISBN 978-7-5635-6106-3　　　　　　　　　　　　　　　　　定　价：19.60 元

前　言

　　本教材共分为两个学习情景：地面服务和客舱服务。其中，地面服务情景有四个学习活动，客舱服务情景有三个学习活动。每个学习活动有相对应的学习任务，每个学习任务分为任务描述、任务准备、任务实施、任务拓展和提升、任务总结和知识点链接，以及课后练习和任务。

　　任务描述部分设定任务角色，设计与工作相关的情景，让学生明确相关任务，便于顺利开展任务。

　　任务准备部分是实施任务前必须掌握的相关知识资料的准备和预习。

　　任务实施部分以学生合作探究的形式进行工作任务的听说读写和角色扮演的活动，以便学生在工作情景中完成相关工作任务，同时附有学生任务完成评价表，便于检验任务完成情况。

　　任务拓展和提升部分以贴近学生实际生活的情景对话的听说读写和相关练习为基本内容，便于教师进行情景教学，同时可提高学生的学习兴趣。每个对话后的核心词汇部分是每个对话所涉及的重点单词和句型，以便让学生明确重点交际单词和句型，增强英语词汇和会话能力。同时，也有相关工作情景的英语广播稿训练，让学生在做中学、学中做，进一步提升学生的综合职业能力。

　　任务总结和知识点链接部分将工作任务的要点和工作流程进行梳理、总结，并明确任务中的相关知识点。

　　课后练习和任务部分是进行任务举一反三的巩固练习，以便检验学习效果。

　　总之，本教材的编写从内容上体现了实用性和专业性，贴近生活、贴近实际、贴近专业；从教学形式上对每个教学环节都做了教学说明，便于教师克服传统英语教学的弊端，开展更有效的、灵活多样的课堂活动，激发学生学习兴趣，使其愉快学习。

目　录

情景一 Ground Service 地面服务

学习活动一 客票服务 (Tickets Service)

任务描述

　　假如你是客票预订中心（Booking Office）服务人员 Annie，现在外国乘客 John 要买一张明天早上北京飞哈尔滨的经济舱机票，请你帮助外国乘客 John 完成咨询、购票以及更改机票的任务。

任务准备

【任务相关知识准备】请根据下面的图片写出正确的英文单词或词组。

_____　　　　_____

　　【小组查询资料并完成讨论】利用电脑、手机等媒介，查询资料并讨论：购票需要了解的票务相关知识有哪些？

任务实施

【小组合作探讨完成下列任务】

写一写：分组讨论撰写"为外国乘客进行客票服务"的对话。

要　求：（1）符合对话逻辑，情节合理；

（2）不少于12句；

（3）语法合乎要求，鼓励创新。

_____

_____

_____

_____

_____

【角色扮演】 为外国乘客进行客票服务。

演一演： 分角色扮演操练（实训场地）。

要　求：（1）可使用自制的小道具进行展示，达到完成服务工作的目的；

（2）在展示过程中注意对话情节合理性和礼貌服务；

（3）每组表演时间不超过5分钟；

（4）2～4人为小组代表在实训区进行展示；

（5）完成小组模拟情境互评打分表。

评价内容	专业能力 70分			职业素养 30分						总分	建议
组别	对话撰写无误，为乘客办理客票服务，流程清晰、完整（30分）		表演中，成员发音清晰，语音语调准确，流程完整，表演流畅（40分）		对话表演积极参与，情绪饱满，动作礼仪规范（15分）			对话表演中有一定的创新性（5分）		对话表演整体效果良好（10分）	100分	
得分	30	20	10	40	30	20	15	10	5	5	3	1	10	7	4		
一组																	
二组																	
三组																	
四组																	
五组																	

任务拓展和提升

【对话】Dialogue A　Purchasing a Ticket　购票

A：Good morning, sir. Can I help you?

P：Yes. I'd like to buy a ticket to Harbin on next Monday.

A：Just a moment, please. I'll check... There are

two flights. The A Airline has a flight leaving at 10:00 a. m., and B Airline has a flight leaving at 4:00p. m.. Which flight would you prefer?

P: The flight is leaving at 10:00 a. m..

A: Thank you. Would you fill out this form, please?

P: OK. Here is the form.

A: Thank you. Well, Mr. Baker, will you fly first class or economy class?

P: Economy class. please.

A: One-way or return?

P: One-way, please. What is the air fare?

A: The fare is 1 480 yuan.

P: Do you have a discount?

A: Yes. 20% discount. 1 184 yuan.

P: All right. Here is 1 200 yuan.

A: May I see your passport, please?

P: OK. here you are.

A: That's fine. Wait a moment, please.

...

A: Here are your passport and ticket. Flight CZ4107, Chengdu to Harbin, next Monday. And your change.

P: Thank you.

A: You're welcome. Have a nice trip.

P: Bye.

Dialogue B Ticket Reissue 更改机票

A: Good morning. A Airline. May I help you?

P: Well, I hope so. I'd like to change my reservation on Flight CZ4107 for next Monday to next Friday.

A: May I see your passport?

P: Yes, here you are.

A: Please wait and let me check...Well, Mr. Baker, would you like to choose the same flight on next Friday?

P: That's fine.

A: Your reservation on the flight CZ4107 for next Monday is cancelled. Instead, I've confirmed the Flight CZ4107 for next Friday. And I'll reissue your ticket.

P: Thank you.

A: My pleasure. OK, here are your ticket and passport. Have a nice trip!

```
┌─────────────────────────────────────────────────────┐
              Key words & expressions

  passport      n. 护照          prefer       vt. 更喜欢

  discount      n. 折扣          change       n. 零钱

  form          n. 表格          instead      adv. 代替，顶替

  reissue       n. 更改机票      choose       vi. 选择，挑选

  pleasure      n. 快乐，希望，娱乐，令人高兴的事

  fill out      填（表格）       flight number    航班号码
└─────────────────────────────────────────────────────┘
```

【巩固练习】

1. 两人一组，分角色朗读对话，注意语音语调。

2. 再次朗读对话，回答下列问题。

1) Who are the two persons in the dialogue A?

_____

2) Where does the passenger want to go in dialogue A?

_____

3) How much is the ticket in dialogue A?

_____

4) How does the receptionist help the passenger in dialogue B?

_____

3. 请听教师的朗读，写出听到的单词或者词组。

1) _____   2) _____

3) _____   4) _____

5) _____   6) _____

7) _____   8) _____

9) _____   10) _____

4. 请根据上下文，填写出合适的句子，使对话完整。

第一篇

A：_____?

P：Yes. I'd like to buy a ticket to Chongqing for tomorrow.

A：Just a moment. _____. The Air China Limited has a flight leaving at 10：00 tomorrow. _____?

P：Good. I'll take that flight.

A：OK. _____, please.

P：OK. Here is the form.

A：_____?

P：First class，please.

A：_____.

P：Yes.

A：_____?

P：Here you are.

A：Thank you. The fare is 500 yuan.

P：OK. Here's 500 yuan.

第二篇

A：(Pick up the phone) Hello，this is _____Booking Office，_____

_____?

P：Yes，I hope so. I'd like to change my reservation on Flight 3U678 from Friday to Thursday.

A：Have you bought a ticket?

P：Yes. _____.

A：I'm afraid you must come to our booking office to reissue your ticket.

P：Really?

A：Yes，we must reissue your ticket.

P：What else should I take with me?

A：_____with you.

P：OK. I see. I'll come later.

A：Thank you.

5. 将下列英文句子翻译成中文或将中文句子翻译成英文。

1) There are two flights. The A Airline has a flight leaving at 10:00 a. m.，and B Airline has a flight leaving at 4:00 p. m.. Which flight would you prefer?

_____

2) Would you fill out this form，please?

_____

3) One-way or return?

_____

4) Do you have a discount?

_____

5) I'd like to change my reservation on Flight CZ4107 for next Monday to next Friday.

_____

6) Please wait and let me check...

_____

7）You must come to our booking office to reissue your ticket.

8）您的护照和机票。CZ4107 下周一从成都飞哈尔滨。还有找您的零钱。

9）祝您旅途愉快！

6. 请大声朗读下列广播稿，注意语音语调。

Ladies and Gentlemen，

    May I have your attention please：

    Flight CA167 from Beijing to Tokyo is now landing.

    Thank you.

任务总结和知识点链接

1. 请总结为乘客订票的流程。

2. 知识点链接。

1）普通票价

普通票价即航空公司公布的销售票价，通常称全价票。全价票在签转、变更和退票等方面的限制很少。

2）特种票价

特种票价通常分为两类，一类适用于政府规定的特定人群，如革命伤残军人和因公致残的人民警察等。此类票在签转、变更和退票等方面的限制很少。另一类是航空公司根据市场淡旺季、购票人数、购票时间的早晚等情况推出的适用于经济舱的票价，通常称折扣客票，购买此种客票时，请务必了解该客票相关的限制条件。

如果购买了特价客票，请一定要仔细了解航空公司的限制条件（一般是不得退票、不得更改、不得签转），以免造成不必要的损失。提前购票能购买到不同折扣的低价客票，飞往同一地方各时段的航班客票价格也有差别，购票时应了解相关信息；此外，客票价格越低，限制条件相对也越多，购票时一定要咨询清楚，以免带来不必要的麻烦。航空公司普遍采用收益管理系统进行机票销售管理，会发生旅客购票时价格随时变化的情况，此属正常现象。

3）头等舱、公务舱票价

为适应国内航空运输市场发展，中国民用航空局、国家发展和改革委员会研究决定，自 2010 年 6 月 1 日起，民航国内航线头等舱、公务舱票价实行市场调节价，由各运输航空公司自主定价。各运输航空公司国内航线头等舱和公务舱价格种类、价格水平及适用条件（含头等舱和公务舱的座位数量、与经济舱的差异以及相匹配的设施和服务标准等）会有所不同。

4）儿童票/婴儿票

满两周岁但不满十二周岁的儿童，按国内航线成人普通票价的50％计价，国际航线根据航线不同而有差异。

未满两周岁的婴儿，使用成人普通票价10％的客票乘机，不单独提供座位；如需要单独占用座位，则应购买儿童票。每位成人旅客只能携带一名婴儿，超过的人数请另行购买儿童票。

儿童、婴儿乘坐国内航班，可以自愿选择购买航空运输企业在政府规定政策范围内确定并公布的其他种类票价，并执行相应的限制条件。但仍免收机场管理建设费。

5）联程客票

如需转乘航班才能到达目的地的旅客，在购买客票时，应查验所购客票是否为联程票，即全部航程均填列在一张或多张票号相连的客票上。

客票有效：如果购买的是普通票价的客票，有效期自实际出发之日起，一年内有效；如果客票未曾使用，则自填开客票之日起，一年内有效。折扣票的有效期，适用航空公司公布的条件，请务必注意。

6）客票有效期的计算

从旅行开始或填开客票之日的次日零时起至有效期满之日的次日零时为止。

课后练习和任务

【课后练习】

1. 请朗读下列句子。

1）提出请求

May I have a look at your ticket?

我可以看看你的机票吗？

Would you show me your tickets please?

可以出示你的机票吗？

May I have your name and telephone number?

我可以知道你的姓名和电话号码吗？

2）征求意见

Which flight would you like/prefer?

你想要哪趟航班？

First class or economy class?

头等舱还是经济舱？

Single or return ticket?

单程票还是往返票？

3）道别

Have a good trip. /Have a good flight.

旅途愉快！

4）道歉

Sorry, sir. All seats for that flight are fully booked.

很抱歉，先生。那架航班的所有座位已经订满。

Sorry, there are no seats left.

对不起，没有剩余的座位。

5）May I ask where you want to go?

请问您想购买到哪里的机票？

6）May I ask when you are going to take the flight?

请问您打算购买几号的机票？

7）There are 4 flights to Beijing on April 25th, departing from Qingdao at 7：10, 11：00, 14：30, and 19：00. The trip takes about 1 hour.

4 月 25 日从青岛到北京的航班有 4 班，起飞时间分别为 7：10、11：00、14：30 和 19：00，空中飞行时间约 1 小时。

8）May I ask whether you want to make a reservation or buy it now?

请问您是想预订还是直接购买？

9）Please fill in the Ticket Purchasing Form.

请您填写购票单。

10）The biggest discount for the flight to Beijing on May 1st is 20%, and it is 600 yuan with the discount.

5 月 1 日到北京的航班折扣最低有 8 折，折扣价为 600 元。

11）Would you like an economy ticket or the first-class ticket?

您想购买普通舱机票还是头等舱机票？

12）Would you like to pay in cash or by credit card?

您想用现金付款还是信用卡付款？

2. 英汉互译，请将下列中文翻译成英文或将英文翻译成中文。

1）direct flight

2）scheduled flight

3）Airport Inquiry Office

4）single ticket

5）make a reservation

6）take off

7）中国东方航空公司

8）归因于

9）往返票

10）填写

11）经济舱

12）头等舱

【课后任务】

分小组，根据下面的背景编写对话，要求英文对话不少于 **12** 句，并将对话录制成视频以小组形式上传到学习通平台。

You want to order a ticket of China Airline flight CA984 from Beijing to Shanghai next Monday through the C-trip on the phone.

任务描述

　　假如你是客票预订中心（Booking Office）服务人员 Annie，外国乘客 Jack 因公司经理安排，临时取消航程，他要退掉一张明天早上从北京飞上海的经济舱机票，请你帮助外国乘客 Jack 完成咨询与退票的任务；同时，他的朋友 William 要订一张下周一从北京飞纽约的机票，请帮助他完成出票的任务。

任务准备

【小组查询资料并完成讨论】利用网络、手机等媒介，查询资料并讨论：

(1) 退机票的流程是什么？

(2) 出一张机票的流程是什么？

任务实施

【小组合作探讨完成下列任务】

写一写：分组讨论撰写"为外国乘客进行退票、出票服务"的对话。

要　求：(1) 符合对话逻辑，情节合理；

　　　　(2) 不少于 12 句；

　　　　(3) 语法合乎要求，鼓励创新。

_____

_____

_____

_____

_____

_____

_____

【角色扮演】退票、出票服务。

演一演：分组角色扮演操练（实训场地）。

要　求：（1）可使用自制的小道具进行展示，达到完成服务工作的目的；

　　　　（2）在展示过程中注意对话情节合理性和礼貌服务；

　　　　（3）每组表演时间不超过 5 分钟；

　　　　（4）2～4 人为小组代表在实训区进行展示；

　　　　（5）完成小组模拟情境互评打分表。

评价内容	专业能力 70分						职业素养 30分									总分	建议
组别	对话撰写无误，为乘客办理"退票、出票"手续流程清晰、完整（30分）			表演中，成员发音清晰，语音语调准确，流程完整，表演流畅（40分）			对话表演中积极参与，情绪饱满，动作礼仪规范（15分）			对话表演有一定的创新性（5分）			对话表演整体效果良好（10分）			100分	
得分	30	20	10	40	30	20	15	10	5	5	3	1	10	7	4		
一组																	
二组																	
三组																	
四组																	
五组																	

任务拓展和提升

【对话】Dialogue A　Ticket Refund　退票

A：Good morning. Can I help you?

P：Well, I want to have my ticket refunded.

A：May I have the flight number and the departure date for the flight?

P：Yes. 3U8751 from Chengdu to Haikou on the 7th of August.

A：May I see your passport, please?

P：Yes, here you are.

A：Thank you. Wait a moment. Your reservation to Haikou is for tomorrow. We can refund your ticket, but you have to pay a certain amount of cancellation fee.

P：I don't quite understand that.

A：Well，according to airline regulations，the passenger who asks for refund 24 hours before flight departure should pay the cancellation fee 10％ of the original fare，and 20％ if refund is requested within 24 hours but before 2 hours ahead of flight departure. The cancellation fee is 30％ of the original fare if it is claimed within 2 hours before departure. Your reservation is for tomorrow. So you should pay 20％ of the 1 480 yuan. That's 296 yuan. The refund is 1 184 yuan.

P：I see.

A：Please fill out the form.

P：That's all right.

A：Here's your passport and your money back.

P：Thank you!

Dialogue B　Issuing Tickets　出票

A：Good morning，Sir.

P：Morning. I'm going to Beijing on Monday. I'd like to know tomorrow's scheduled flight for Beijing.

A：We have two flights for Beijing tomorrow，CZ3151，leaving at 8：00 in the morning and CA1304 leaving at 20：00 in the evening.

P：CZ3151，please.

A：OK. Would you fill out this form，please? Your name，flight number，route，date，contact telephone number and so on.

P：Here you are. How much is it?

A：Excuse me，do you want single or return ticket?

P：Single ticket，please.

A：RMB 1 000. Yes，good-bye. Have a good trip!

Key words & expressions

refund	*n.* 退（票）；*vt.* 退（票）	regulation	*n.* 规定，规则
request	*v.* 要求	visa	*n.* 签证
original	*adj.* 原来的，原始的	amount	*n.* 数量
fee	*n.* 费用，小费	issue	*v.* 发出，发行
urgent	*adj.* 紧急的，急迫的	contact	*v.* 联系
issue tickets	出票	fill in＝fill out	填写
single ticket	单程票	return ticket	往返票
non-stop flight	直达航班		

【巩固练习】

1. 两人一组，分角色朗读对话，注意语音语调。

2. 再次朗读对话，回答下列问题。

1）Who are the two persons in the dialogue?

_____

2）What's the matter with the passenger?

_____

3）What's the reservation of the passenger?

_____

4）What's the regulation of the passenger who asked for refund?

pay the cancellation fee 10％ of the original fare：_____

pay the cancellation fee 20％ of the original fare：_____

pay the cancellation fee 30％ of the original fare：_____

5）Who are the two persons in the dialogue?

_____

6）When will the passenger fly the CZ3151?

_____

7）How much is the ticket for CZ3151?

_____

3. 请听教师的朗读，写出听到的单词或者词组。

1）_____　　2）_____

3）_____　　4）_____

5）_____　　6）_____

7）_____　　8）_____

9）_____　　10）_____

4. 将下列英文句子翻译成中文或将中文句子翻译成英文。

1）Due to the bad weather conditions，this flight will not arrive on schedule.

_____

2）Could I make two economy class reservations for this flight?

_____

3）I'd like to change the Qingdao-Beijing reservation and reserve a seat to Tianjin.

_____

4）Please be at the airport 2 hours before your flight time for check-in.

_____

5) I would like to apologize for the inconvenience caused by this delay.

_____

6) 有什么需要帮忙的吗？

_____

7) 我们每周二和周四都有航班飞往武汉。

_____

8) 本次航班经停济南，总航程需要 2 小时。

_____

9) 您想购买普通舱机票还是头等舱机票？

_____

10) 祝您旅途愉快！

_____

5. 请大声朗读下面的广播稿，注意语音语调。

Passengers taking supplementary flight CA167 from Beijing to Tokyo，attention please：

Please go to the Counter/Service Counter/Information desk No. 8 to exchange your boarding passes for transit passes.

Thank you.

任务总结和知识点链接

1. 为乘客退票流程是什么？

2. 为乘客出票流程是什么？

3. 知识点链接。

1) 客票有效期：如果购买的是普通票价的客票，有效期自实际出发之日起，一年内有效；如果客票未曾使用，则自填开客票之日起，一年内有效。折扣票的有效期，适用航空公司公布的条件，请务必注意。

客票有效期的计算：从旅程开始或填开客票之日的次日零时起至有效期满之日的次日零时为止。

2）客票变更：如果在购票后要求改变航班、日期、舱位等级等，属于自愿变更。自愿变更有一定限制，折扣票有不同的限制条件，甚至不允许变更。

3）退票：如果在客票有效期内要求退票，凭有效身份证件，按航空公司的规定即可办理。电子客票凭行程单办理退票手续，如果未曾打印行程单，可凭有效身份证件办理退票手续。如果在航班经停地自动终止旅行，则该航班未使用航段的票款不退。如果因病要求退票，需提供医疗单位证明，始发地退还全部票款，经停地退还未使用航段的全部票款，不收取手续费。患病旅客的陪伴人员要求退票，不收取手续费。

票款只能退给客票上列明的旅客本人。如委托他人办理，须提供旅客本人和受委托人的有效身份证件。如果持折扣客票，退票按航空公司的有关规定办理。

4）客票遗失：当遗失客票时，请以书面形式迅速向航空公司或其销售代理人申请挂失。申请挂失前，客票如果已被冒用或冒退，航空公司不承担责任。申请挂失后，经查证客票未被冒用或冒退，待客票有效期（通常是一年）满后的 30 天内，办理退款手续。国际客票应在航空公司规定的时限内办理。

5）航空保险：

客票险——您的客票含有旅客法定责任险。当您在乘坐飞机发生意外时，造成人身伤亡或行李物品的损失均属法定责任险。

航意险——您可以自愿向航空公司投保国内航空运输意外伤害险。购买此险，不免除或减少航空公司应当承担的赔偿限额。

课后练习和任务

【课后练习】
1. 请将下列英文短语翻译成中文。

1）have a refund for one's ticket

_____

2）change flight and date

_____

3）pay for the refund

_____

4）the least amount

_____

5）ask for a refund

_____

6）original fare

_____

7）before the departure time

_____

8）reschedule one's ticket

_____

9）miss flight

_____

2. 请根据上下文，按照中文填写出合适的句子，使对话完整。

（The Olympic swimming coach Mr. Gilbert has missed the flight CA982 to Paris. He goes to the Air China Ticket Office to reschedule his ticket. ）

C：_____ （您好，有什么事吗？）

P：I just missed my flight due to a traffic jam. Can I reschedule it?

C：_____ （可以，您要去哪儿？）

P：Paris.

C：_____ （CA1304 后天上午 11：40 起飞。）

P：Do you have anything earlier?

C：Let me check. _____ （CA542 明天下午 15：25 起飞，还有空座，改签这趟航班行吗？）

P：OK. I'll take that.

C：Let me have your passport and your original ticket，please.

P：_____ （给您，顺便问一下，我还要再付费吗？）

C：No, you don't. Here are your ticket and passport.

P：Thank you.

C：_____ （感谢您选择国航的航班！）

【课后任务】

分小组，根据下面的背景编写对话，要求英文对话不少于 **12** 句，并将对话录制成视频以小组形式上传到学习通平台。

The Olympic runner Mr. Robert has bought a ticket on flight CA321 to Washington. But he has to cancel his reservation and asks for a refund of his ticket 15 hours before the departure time because the Olympic program for events has changed.

_____

_____

_____

_____

_____

_____

_____

_____

_____

情景一　Ground Service 地面服务
学习活动二　值机服务（Check-in Service）

假如你是值机柜台的地面（Check-in Counter）服务人员 Annie，外国乘客 Mike 现在正在办理 CA901 次航班从北京飞往纽约的值机手续，请你帮助外国乘客 Mike 完成咨询、值机的任务。

任务准备

【任务相关知识准备】请看图片，依据中文在横线上填写相应的英文。

登机牌

_____

电子客票行程单

_____

身份证

_____

行李

_____

前面的

_____

后面的

_____

行李交运单

_____

【小组查询资料并完成讨论】利用电脑、手机等媒介，查询资料并讨论：办理值机手续的流程是怎样的？

任务实施

【梳理句型】小组合作完成下列句型。

1. 探究学习：听音频，并将下列对话中缺失的句子补充完整并填在横线上。

Sense：Mr Chen is going to take Flight CA3603 to Macao. It's time to check-in. He goes to the Air China check-in counter.

C＝Clerk（工作人员）；P＝passenger（乘客）

C：Good afternoon.

P：Good afternoon. _____.

C：Yes. _____?

P：Here you are.

C：_____?

P：Only a hand bag.

C：All right. _____?

P：Please give me a window seat.

C：All right. _____.

_____. Have a nice trip.

2. 读一读：组内分角色朗读上面的对话。

3. 学习通：请完成上述对话的配音练习，注意语音语调。

【明确办理值机手续流程】

1. 听对话音频，按要求完成对话的相关单词和句子填空（填充重点句子）。

1）整理出本对话中机场地面服务人员问候、提出帮忙的常用语。

_____

_____

_____

2）根据以上内容，将对话内容涉及的句子填写在下列表格内。

角色		乘客	地勤人员
办理乘机手续的流程	航班号 目的地		
	出示证件 电子客票		
	选座位		
	行李托运		
	换好登机牌		

2. 分组发言：想一想、议一议办理值机的步骤是什么？

办理乘机手续单词要点提示图如下：

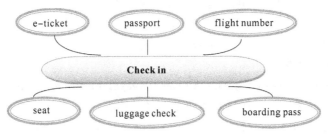

【小组合作探讨完成下列任务】

写一写：分组讨论撰写"为乘客 Mike 在值机柜台办理乘机手续"的对话。

要　求：（1）符合办理乘机手续的流程；

　　　　　　（2）要求使用本课呈现过的句型，不少于15句；

　　　　　　（3）语法合乎要求，鼓励创新。

可参考资料：机舱分布图（以波音 787 机型为例）

布局图：

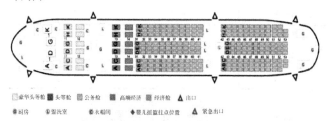

□乘华头等舱 ■头等舱 ■公务舱 ■高端经济 ■经济舱 △山口

●厨房 ●盥洗室 ●衣帽间 ◆婴儿摇篮挂点位置 △紧急出口

可参考语句：

1）Could you put on the luggage label，please?

请您将行李放在行李传送带上好吗?

2）Hand-luggage is not be weighed.

手提行李不用称重。

3）Would you put your baggage on the conveyor belt one by one，please?

请把您的行李一件一件放在传送带上好吗?

4）I want to have a window seat.

我想要一个靠窗的位置。

5）Here's my passport. I'm going on the CA Flight to New York.

这是我的护照，我将乘坐国航的航班到纽约。

6）May I have your passport and Itinerary of e-ticket.

请出示你的护照和行程单。

7）Could I take this briefcase as my hand baggage?

这个公文包能作为手提行李吗?

8）What kind of seat do you prefer?

您喜欢什么座位?

9）I'm sorry. No more aisle seats are available. Will the window seat be all right?

很抱歉，您要的座位没有了，给您推荐一个靠窗的座位，可以吗?

【角色扮演】为外国乘客 **Mike** 在值机柜台办理乘机手续。

演一演：分角色扮演操练（实训场地）。

要　求：（1）可使用自制的小道具进行展示，达到完成服务工作的目的；

　　　　（2）在展示过程中注意办理乘机手续的流程和礼貌服务；

　　　　（3）每组表演时间不超过 5 分钟；

　　　　（4）2～4 人为小组代表在实训区进行展示；

　　　　（5）完成小组模拟情境互评打分表。

评价内容	专业能力 70 分			职业素养 30 分			总分	
组别	对话撰写无误，为乘客办理乘机手续流程清晰完整（30分）	表演中，成员发音清晰，语音语调准确，流程完整，表演流畅（40分）	对话表演中积极参与，情绪饱满，动作礼仪规范（15分）	对话表演有一定的创新性（5分）	对话表演整体效果良好（10分）	100 分	建议
得分	30 20 10	40 30 20	15 10 5	5 3 1	10 7 4		
一组							
二组							
三组							
四组							
五组							

任务拓展和提升

【知识点】

1. 单词

flight number passport check in counter rear aisle boarding pass luggage check e-ticket

2. 句型

乘客角色

1）说明值机航班号和目的地：Can I check in here for 航班号 to 地点.

2）说明自己心仪的座位：Please give me a 靠窗、过道（window/aisle）seat.

工作人员角色

1）要求乘客出示相关证件和电子客票：Will you give me your 证件/电子客票/护照?

2）询问行李托运事宜：How many pieces of luggage（行李）do you want to check, sir? Please put your baggage checks on your baggage.

3）询问选座位：What kind of seat do you prefer?

4）换好登机牌：Here's your 证件和登机牌（ID card and your boarding pass.）

【技能点】 办理乘机手续流程的操作要点

证件/电子客票/航班信息→选座位→行李托运→换好登机牌

【课后作业】

分组上传"在值机柜台为乘客办理乘机手续"的规范对话表演视频到"学习通"平台。

任务描述

　　假如你是值机柜台（Check-in Counter）的地面服务人员 Annie，外国乘客 Joe 是一位坐轮椅的乘客，现在需要你帮助为其办理 CA901 次航班从北京飞往纽约的值机手续，请你帮助他完成咨询、值机的任务。

任务准备

【知识准备】小组查询资料并完成讨论，写出下列乘客的具体含义。

无人陪伴儿童：

行动不便乘客：

素食乘客：

孕妇：

残疾人：

【小组查询资料并完成讨论】为特殊乘客办理乘机手续的注意事项有哪些？

任务实施

【对话】特殊旅客办理乘机流程

Dialogue A   unaccompanied minor   无人陪伴儿童

C＝Clerk（工作人员）；P＝Passenger（乘客）

C：Good morning. May I help you?

P：Yes. I am here for the boy. He will check-in.

C: Well, he is Li Hua, the unaccompanied minor.

P: Yes. I'm his teacher. He will go to Tibet to meet his parents during his summer holiday.

C: I'd like to let you know that we'll seat Li Hua a forward row so that he will be easy for our flight attendants to find and watch.

P: You are so thoughtful, thank you so much.

C: We have prepared an envelope to contain his travel documents. As Li Hua's guardian, could you please read carefully and complete the form on the cover of the envelope?

P: Here you are. Is that all right?

C: Yes. Thank you. This is your copy. I'll put his travel documents in the envelope and pass it to our senior flight attendant on his flight.

P: Be sure that your flight attendants will watch him on board.

C: Yes. Here is a badge for an unaccompanied child.

(To Li Hua) Let me pin this badge on your chest so that our flight attendants can identify you and look after you during the flight.

(To the guardian) He will soon be escorted to the cabin crew.

P: Is there anyone to assist him after descending at Lhasa Airport?

C: Yes, our ground staff at the airport will be informed and ready to meet him.

P: Thank you very much.

Dialogue B Missing the Time for Check-in 错过值机时间

P: It's the counter of China Southern, isn't it?

A: Yes. What can I do for you?

P: Can I check in for flight CZ3102 to Guangzhou.

A: Sorry, sir. You are late for check-in. The system has been closed off. You have to go to the duty managers'counter. The agent there will help you get a boarding pass.

...

(At the duty managers' counter)

P: Excuse me, can you help me?

A: Sure. What's the problem?

P: You see, I've missed check-in time for flight CZ3102. I wonder if you could help me check in that flight.

A: Can I have your passport, please?

P: Here you are.

A: Do you have any baggage to check-in?

P: No. I only have a briefcase.

A：OK．Just a moment．Here is your boarding pass．Your seat is 33E．The passengers are expected to start boarding in ten minutes.

P：Where should I go through the security check?

A：Just over there behind that counter．You should go to gate 11 as soon as you finish security check.

P：Thank you very much.

Key words & expressions

badge	*n.* 徽章，像章	document	*n.* 文献，文件；公文
escort	*v.* 护送，伴随	guardian	*n.* 监护人
identify	*v.* 辨认，识别	lounge	*n.* 休息室
minor	*n.* 未成年者	pin	*v.* （用钉等）钉住，别住
staff	*n.* （全体）职员、工作人员		
thoughtful	*adj.* 体贴人的，对……关心的		
unaccompanied	*adj.* 无人伴随的		
system	*n.* 制度，体制；系统；方法		
counter	*n.* 柜台；对立面；计数器	problem	*n.* 难题；引起麻烦的人
miss	*vt.* 漏掉；错过；想念	briefcase	*n.* 公文包，公事包
security	*n.* 安全	China Southern	中国南方航空

【小组合作探讨完成下列任务】

写一写：分组讨论撰写"为轮椅乘客 Joe 在值机柜台办理乘机手续"的对话。

要　求：(1) 符合办理乘机手续的流程；

(2) 要求使用本课呈现过的句型，不少于15句；

(3) 语法合乎要求，鼓励创新。

_____

_____

_____

_____

_____

【角色扮演】为轮椅乘客 Joe 在值机柜台办理乘机手续。

演一演：分角色扮演操练（实训场地）。

要　求：(1) 可使用自制的小道具进行展示，达到完成服务工作的目的；

(2) 在展示过程中注意办理乘机手续的流程和礼貌服务；

(3) 每组表演时间不超过5分钟；

(4) 2～4人为小组代表在实训区进行展示；

(5) 完成小组模拟情境互评打分表。

评价内容	专业能力 70 分						职业素养 30 分									总分	建议
组别	对话撰写无误，为乘客办理乘机手续流程清晰、完整（30 分）			表演中，成员发音清晰，语音语调准确，流程完整，表演流畅（40 分）			对话表演中积极参与，情绪饱满，动作礼仪规范（15 分）			对话表演有一定的创新性（5 分）			对话表演整体效果良好（10 分）			100 分	
得分	30	20	10	40	30	20	15	10	5	5	3	1	10	7	4		
一组																	
二组																	
三组																	
四组																	
五组																	

【巩固练习】

1. 分角色朗读上面的对话。

2. 根据对话 A 回答下列问题。

1）Who are the two persons in the dialogue?

_____

2）What's the matter with the passenger in dialogue B?

_____

3）How did the attendant help the unaccompanied minor in dialogue A?

_____

3. 根据对话 B 回答下列问题。

1）Who are the two persons in the dialogue?

_____

2）What's the matter with the passenger in dialogue B?

_____

3）How did the attendant help the missing time in check passenger in dialogue B?

_____

4. 听教师朗读单词或句子，将所听内容写在下面的横线上。

1）_____   2）_____

3）_____   4）_____

5）_____   6）_____

7）_____   8）_____

9) _____   10) _____

5. 将下列英文短语翻译成中文并记忆。

1）unaccompanied minor   2）VIP

_____   _____

3）travel documents   4）pin this badge

_____   _____

5）cabin crew   6）ground staff

_____   _____

7）see somebody. off   8）boarding gate

_____   _____

6. 将下列英文句子翻译成中文或将中文翻译成英文。

1）We expect to start boarding at 4:30 p. m..

_____

2）The passengers are expected to start boarding in ten minutes.

_____

3）您错过办理登机的时间了。

_____

4）现在系统已经关闭了。您要去值班经理柜台办理。

_____

5）You should go to gate 11 as soon as you finish security cheek.

_____

任务总结和知识点链接

1. 小组讨论并总结完成为特殊旅客办理值机手续的流程。

2. 知识点链接——关于特殊旅客的相关知识点。

重要旅客的售票手续办理后，应在起飞前24小时发特殊旅客通知（SPA/VIP）起飞前一天报机场要客服务。重要旅客取消旅行或改变乘机地点、日期，办理变更的售票处应及时发报通知有关空运企业和办事处。

儿童：（1）定义：2～12周岁的儿童。

（2）分类：①有成人陪伴的儿童（包括婴儿），未满2周岁，10%不占座位。

②无成人陪伴的儿童：无成人陪伴儿童指5周岁至12周岁以下无成人陪伴，单独乘机的儿童，5周岁以下的无成人陪伴儿

童，不予承运。无人陪伴儿童必须符合航空公司的运输条件方可运输。

病残旅客：（1）定义：由于身体和精神上的病态，空中旅行中不能照顾自己，需要由他人照料的旅客，称为病残旅客。带有先天残疾，已习惯于自己生活的人不应视为病残旅客。

（2）接受病残旅客：①诊断证明书；②填写特殊旅客（病残）乘机申请书；③陪伴人员。

（a. 病残旅客乘机，原则上由医生或护理人员陪同，以便旅途中照料旅客。b. 除精神病人外，符合条件的可单独旅行。）

担架：（1）担架旅客订座不得迟于航班起飞前72小时，每一航班每一航段上只限载运一名担架旅客。

（2）陪伴人员经医生同意，至少一名陪护人员。

孕妇：（1）怀孕32周及以下的孕妇乘机，除不适宜乘机者外，可按一般旅客运输。

（2）怀孕超过32周的孕妇，应提供医生诊断证明。

（3）怀孕超过35周（含）的孕妇，预产期4周（含）以内者，一般不予接受运输。

（4）起飞前48小时，拍发特殊旅客运输通知SPA/PREGNANT电报发给经停站和到达站。

课后练习和任务

【课后练习】

1. 完成下列对话。

C：Excuse me, madam. Are you Cairo's mother?

P：No. I am her aunt.

C：I'm sorry. I'd like to let you know _____ so that she will be easy for our night attendants to find and watch.

P：_____.

P：I'll put her travel documents in an envelope and pass it to _____

P：Be sure that your flight attendants will watch him on board.

C：Let me _____ on her chest so that our flight attendants can identify you and look after you during the flight.

P：Is there anyone to assist him after landing at Lhasa Airport?

C：Yes, our ground staff at the airport will be informed and ready to meet him.

P：Thank you very much.

2. 用下列单词或短语完成对话。

the cover; travel documents; an unaccompanied child; an envelope; forward;

the destination airport

P：I requested special service for _____ when I booked the ticket.

A：Can I have a look at his _____?

P：Of course，here you are.

A：OK，sir. Your son's _____ are in order. We'll seat him in a _____ _____ row so that it will be easy for our flight attendants to find and look after him.

P：It's very considerate of you. Thank you.

A：Who will come to meet him at _____?

P：His uncle.

A：We have prepared _____ to contain his travel documents. Could you please complete this form on _____ of the envelope?

P：Certainly... Here you are. Is that all right?

3. 请根据下列句子的提示补全句子，使句子完整。

P：Good afternoon. I'm going to Brussels. Shall I check in at your counter?

A：_____?

P：Good morning. Is this the check-in counter for Flight CA5513 to Hangzhou?

A：_____.

A：_____?

P：I'd like an aisle seat.

A：_____?

P：Two small suitcases.

A：_____?

P：Yes，a handbag.

P：Can I have window seat in the forward section?

A：I'm afraid _____.

P：Can I check in for flight CZ3102 to Guangzhou.

A：Sorry, sir. You are late _____.

P：Is there anyone to assist him through Immigration and Customs after landing at Manila Airport?

A：Yes, our ground staff _____.

P：I think I'll need your special assistance.

A：_____（为您安排轮椅）.

4. 根据下列句子填出空缺的单词或短语。

1）You are _____ check-in.

2）You have to go to the duty _____ counter.

3）The system has been closed off.

4）Do you have any baggage to _____ ?

5）The passengers are _____ to start boarding in ten minutes.

5. 将下列英文句子翻译成中文。

1）It's the counter of China Southern，isn't it?

_____

2）The system has been closed off. You have to go to the duty managers' counter.

_____

3）Where should I go through the security check?

_____

4）Just over there behind that counter.

_____

5）I wonder if you could help me check in that flight.

_____

【课后任务】

分小组，根据下面的背景编写对话，要求英文对话不少于 **12** 句，并将对话录制成视频以小组形式上传到学习通平台。

A woman with a baby, taking a big suitcase is checking in on the check-in Counter，please help her to check-in.

_____

_____

_____

_____

_____

_____

_____

_____

为特殊旅客办理值机子任务结束

情景一　Ground Service 地面服务

学习活动三　行李服务（Baggage Service）

　　假如你是自助行李值机柜台的工作人员 Rose，请你帮助外国乘客 Amy 办理一件重约 15 kg 行李箱的自助行李托运。

任务准备

　　【小组查询资料并完成讨论】利用网络、手机等媒介，查询资料并讨论：

　　（1）为乘客办理行李托运必须了解的知识点——了解行李托运规定的相关信息。请将之前搜集整理的航空公司的免费行李额度写出来。

　　（2）询问行李件数以及行李称重的相关英语常用表达。

任务实施

　　【小组合作探讨完成下列任务】

写一写：分组讨论撰写"为乘客 Rose 在行李值机处办理行李托运手续"的对话。

要　　求：（1）符合办理行李托运手续的流程，不少于 12 句；

　　　　　　（2）语法合乎要求，鼓励创新。

_____

_____

_____

_____

【角色扮演】为外国乘客 Rose 在行李值机处办理行李托运手续。

演一演：分角色扮演操练（实训场地）。

要　求：（1）可使用自制的小道具进行展示，达到完成服务工作的目的；

　　　　　（2）在展示过程中注意办理流程和礼貌服务；

　　　　　（3）每组表演时间不超过 5 分钟；

　　　　　（4）2～4 人为小组代表在实训区进行展示；

　　　　　（5）完成小组模拟情境互评打分表。

评价内容	专业能力 70 分			职业素养 30 分									总分	
组别	对话撰写无误，为乘客办理行李托运手续流程清晰、完整（30分）			表演中，成员发音清晰，语音语调准确，流程完整，表演流畅（40分）			对话表演中积极参与，情绪饱满，动作礼仪规范（15分）			对话表演有一定的创新性（5分）			对话表演整体效果良好（10分）			100分	建议
得分	30	20	10	40	30	20	15	10	5	5	3	1	10	7	4		
一组																	
二组																	
三组																	
四组																	
五组																	

任务拓展和提升

【对话】朗读并注意画横线的句子。

A：Hello，Sir.

B：Hello，is this <u>where I ought to have my baggage checked</u>?

A：Yes，have you gone through all the formalities of customs，quarantine and frontier inspection?

B：Yes，I have. <u>What is the allowance for each passenger traveling on an international flight</u>?

A：On international flights，<u>the free baggage allowance for each passengers：first class，40 kg；business class，30 kg；economy class，20 kg. No free baggage allowance</u>

is granted to infant paying 10% of the adult fare.

B：That's my baggage. Three pieces of baggage altogether.

A：<u>Let me weigh them on the scales.</u> Well，it's 45 kg altogether，5 kg overweight.

B：50 yuan. Please pay me here.

A：Thanks，bye-bye!

B：Bye-bye.

Key words & expressions

formality	*n.* 正式手续	customs	*n.* 海关
quarantine	*n.* 检疫	allowance	*n.* 允许
weigh	*v.* 称重	fare	*n.* 费用
scale	*n.* 磅秤	overweight	*adj.* 超重的
frontier inspection	边检	be granted to	给予

【巩固练习】

1. 分角色朗读上面的对话。

2. 朗读对话，并回答下列问题。

1）What's the two roles in the dialogue?

_____

2）According to the dialogue，what's the allowance for each passenger traveling on an international flight?

_____

3）What's the problem with the passenger's baggage?

_____

3. 听老师朗读，写出听到的单词或句子。

1）_____   2）_____

3）_____   4）_____

5）_____   6）_____

7）_____   8）_____

9）_____   10）_____

4. 请将下列中文句子翻译成英文。

1）您想要什么样的座位？

_____

2）您的行李一共 10 千克。

_____

3）航空公司规定每人手提行李只准带一件，限重 5 千克。

_____

4）按规定，您的免费托运行李额重为 20 千克。

_____

5）很抱歉，您的行李超重 5 千克。

_____

任务总结和知识点链接

1. Checked Baggage Service **中的交际要点。**

1）说明行李规定：

2）询问行李数量：

3）行李称重：

4）告知行李重量：

5）粘贴行李标签：

2. 知识点链接——托运、取行李流程。

1）每位旅客的免费行李额（包括托运和自理行李）。

2）持成人或儿童票的头等舱旅客为 40 千克，公务舱旅客为 30 千克，经济舱旅客为 20 千克。

3）找到柜台直接把证件（身份证、护照等）交予办理登机手续的人员，托运的行李放在行李传送带上称重，将行李过检，如果行李安全，工作人员会将托运好的行李标签贴在行李上，并将标签存根给乘客，以便核对行李信息。

4）下机后，找到对应的航班行李信息，前往取行李处，对好相应的行李标签，提取自己的行李。

课后练习和任务

【课后练习】

1. 熟读对话，并分角色朗读。

2. 根据上下文，填写句子，使对话完整。

A：Good afternoon. _____.

P：Here you are. I'm going on the CA Flight to Beijing.

A：_____.

P：I have a trunk and a handbag.

A：All right. _____. Thank you for flying CA Airline.

C：Good morning, Mr. Li. I see _____. However, there has

been a change of aircraft for that flight which affects the seat number.

P：Can I get the similar seat in the changed aircraft?

C：I'm sorry. There aren't _____ in the forward section.

P：It's a pity. Can you give me a window seat in the middle or rear cabin?

C：Please wait for a while，I'll check. Yes. I offer _____

____. And if a window seat in the forward becomes available at the last minute，our agent will contact you in the cabin. Is that all right for you?

P：That's OK.

C：Thank you，Mr. Zhao. We're very sorry to cause you this problem.

【课后任务】

　　分小组，根据下面的背景编写对话，要求英文对话不少于12句，并将对话录制成视频以小组形式上传到学习通平台。

　　CA169北京飞往大阪的航班已经开始进行值机服务，请地面服务工作人员帮助经济舱乘客办理1件20千克的行李箱托运任务。

A：_____

P：_____

A：_____

P：_____

A：_____

P：_____

A：_____

P：_____

A：_____

P：_____

A：_____

P：_____

任务描述

　　假如你是"超规行李补托运柜台"的服务人员 Jane，请你帮助外国乘客 A 办理超大行李托运（两个纸箱且长宽高超过 200 cm）、乘客 B 办理超重行李（经济舱 40 kg 的行李箱）的托运任务。

任务准备

【小组查询资料并完成讨论】 利用网络、手机等媒介，查询资料并讨论：

(1) 免费行李额：

(2) 超大行李：

(3) 超重行李：

(4) 办理行李托运流程：

任务实施

【小组合作探讨完成下列任务】

　　写一写：分组讨论撰写"乘客 A 办理超大行李托运、乘客 B 办理超重行李托运手续"的对话。

要　求：（1）符合办理行李托运手续的流程，不少于 12 句；

　　　　（2）语法合乎要求，鼓励创新。

_____

_____

_____

_____

_____

_____

_____

【角色扮演】为"乘客 A 办理超大行李托运、乘客 B 办理超重行李托运手续"的对话。

演一演：分角色扮演操练（实训场地）。

要　求：（1）可使用自制的小道具进行展示，达到完成服务工作的目的；

　　　　（2）在展示过程中注意办理流程和礼貌服务；

　　　　（3）每组表演时间不超过 5 分钟；

　　　　（4）2～4 人为小组代表在实训区进行展示；

　　　　（5）完成小组模拟情境互评打分表。

评价内容	专业能力 70 分						职业素养 30 分									总分	建议
组别	对话撰写无误，为乘客办理超规行李托运手续流程清晰、完整（30分）			表演中，成员发音清晰，语音语调准确，流程完整，表演流畅（40分）			对话表演中积极参与，情绪饱满，动作礼仪规范（15分）			对话表演有一定的创新性（5分）			对话表演整体效果良好（10分）			100分	
得分	30	20	10	40	30	20	15	10	5	5	3	1	10	7	4		
一组																	
二组																	
三组																	
四组																	
五组																	

任务拓展和提升

【对话】Dialogue A　Excess Baggage 超重行李

C：Good morning. May I help you?

P：Good morning. Is this the counter for CA986?

C：Yes. Can I have your passport and itinerary of e-ticket，please?

P：Here you are.

C：What baggage are you going to check in?

P：I have two boxes.

C：It seems that one box is too big. Have you measured it?

P：I haven't measured it. But I weighed them. They are within the free baggage allowance.

C：I don't mean they are overweight. I mean one of your baggage is too long.

P：Are you sure about that?

C：Let me measure. The box is 95 cm long，60 cm wide and 50 cm high. The other one is 60 cm long，40 cm wide and 26 cm high. The big one is over the length limit.

P：I don't understand.

C：The free baggage allowance for each economy passenger is two pieces，and the sum of the length，width and height of each must not exceed 158 cm. but the sum of the two piece must not exceed 273 cm.

P：I see. What should I do then?

C：Yon can send the big box as unaccompanied baggage.

P：Oh，no. I don't want to do that. I want all my baggage to travel with me.

C：I'm sorry. We have to treat your baggage as overweight baggage.

P：That's OK.

C：You have to pay 21 yuan as the excess charge.

P：Here is the money.

C：Here are your boarding pass，and your baggage checks and receipt for the overweight baggage.

P：Thank you.

Dialogue B　Oversized Baggage 超大行李

P：I'm going to take flight CA101 to Hong Kong. Shall I check in at this counter?

A：Yes. Your passport, please.

P：Here you are.

A：Have you got any baggage to check?

P：Yes，these two cardboard boxes.

A：The small one is OK. But the other one is too big.

P：But they only weigh 19 kilograms altogether.

A：I don't mean they are overweight. I mean that one is too long. It's 120 cm long，over the length limitation.

P：What's the limitation?

A：The length，width and height mustn't exceed 100 cm，60 cm，and 40 cm respectively.

P：What should I do then?

A：The big box should be handled as overweight baggage. I'm afraid you have to pay for the excess.

P：OK. Can I pay here?

A：No. There's an excess baggage desk over there. After you pay，please come back with the receipt for the excess. Then I'll check the baggage for you.

P：OK. Thank you.

Key words & expressions

sum	*n.* 总数，总计，总额	excess	*adj.* 过量的，超过限额的
kilogram	*n.* 千克	measure	*adj.* 量，计量
overweight	*adj.* 超重的	receipt	*n.* 收据
cardboard	*n.* 硬纸板；纸板箱	limitation	*n.* 限度；有效期限；
length	*n.* 长度，长	width	*n.* 宽度；广度
height	*n.* 高度；身高；顶点	exceed	*vt.* 超过；胜过
respectively	*adv.* 分别地；各自地，独自地		

【巩固练习】

1. 分角色朗读上面的对话。

2. 根据对话 A，回答下列问题。

1）Who are the two persons in the dialogue?

_____

2）What's the matter with the passenger?

_____

3）What's the regulation of free baggage allowance for each economy passenger?

_____

3. 根据对话 B，回答下列问题。

1）Who are the two persons in the dialogue?

_____

2）Which flight did the passenger want to take?

_____

3）What's the limitation of free baggage allowance?

_____

4. 听老师朗读并写出听到的单词或句子。

1) _____   2) _____

3) _____   4) _____

5) _____   6) _____

7) _____   8) _____

9) _____   10) _____

5. 在横线上填写合适的句子并将对话翻译成中文。

C：Good morning. May I help you?

P：Good morning. Is this the counter for CA301?

C：Yes. Can I have your passport and itinerary of e-ticket，please?

P：Here you are.

C：What baggage are you going to check in?

P：_____.

C：It seems that one box is too big.

P：Really?

C：I think they are overweight.

P：Are you sure about that?

C：Let me weigh. _____.

P：I see. What should I do then?

C：I'm sorry. We have to _____ as overweight baggage.

P：That's OK.

C：You have to pay 21 yuan as the excess charge.

P：Here is the money.

C：Here are your boarding pass，and your baggage checks and receipt for the overweight baggage.

P：Thank you.

C：Let me weight.

任务总结和知识点链接

1. 超大超重行李定义：

2. 超大超重行李办理托运的流程：

3. 知识点链接——航空公司关于行李的知识点。

1）除了随身携带物品之外都属于自理行李，背包如果随身携带就属于随身携带物品，如果办托运就属于托运行李。

2）免费行李额：持成人票或儿童票的旅客，每位免费行李额为：头等舱 40 kg，公务舱 30 kg，经济舱 20 kg。持婴儿票的旅客，无免费行李额。

3）逾重行李额：旅客对超过免费行李额的行李应支付逾重行李费。逾重行李费率以每公斤按经济舱票的 1.5% 计算，金额以元为单位，尾数四舍五入。

4）随身携带物品：随身携带物品的重量，每位旅客以 5 kg 为限。持头等舱客票的旅客，每人可携带两件物品，持公务舱或经济舱客票的旅客，每人只能随身携带一件物品。每件随身携带物品的体积均不得超过 20×40×55 cm。超过上述重量、件数或体积限制的随身携带物品，应作为托运行李托运。

5）不准作为行李运输的物品：国家规定的禁运物品、限制运输物品、危险物品包括但不限于液体、压缩气体、易燃、腐蚀流体、易燃流体和固体、毒品、氧化剂、可聚合物质、磁性物质、放射物质、有害或有刺激性物质，以及具有异味或容易污损飞机的其他物品，不能作为托运行李或随身携带物品。

6）不准在托运行李内夹带的物品：旅客不得在托运行李内夹带重要文件、资料、外交信袋、证券、货币、汇票、贵重物品、易碎易腐物品，以及其他需要专人照管的物品。

7）行李包装：托运行李必须包装完善、锁扣完好、捆扎牢固，能承受一定的压力。对包装不符合的行李，可拒绝承运。

8）行李赔偿：旅客的托运行李从托运时起到交付时止，如发生丢失或损坏，航空公司应承担责任，赔偿金额每公斤不超过人民币 50 元。如行李价值每公斤低于 50 元时，按实际价值赔偿。行李赔偿的重量按旅客实际托运行李重量计算。

免费行李额分为计重免费行李额和计件免费行李额两种。计重免费行李额：按照旅客所付的票价座位等级，每一全票或半票旅客免费行李额为：一等舱为 40 kg，公务舱为 30 kg，经济舱（包括旅游折扣）为 20 kg，按成人全票价百分之十购票的婴儿无免费行李额。计件免费行李额，按照旅客所付的票价座位等级，每一全票或半票旅客的免费行李额为两件，每件长、宽、高三边之和不得超过 158 cm，每件重量不得超过 32 kg。但持有经济舱（包括旅游折扣）客票的旅客，其两件行李长、宽、高的总和不得超过 273 cm，按成人全票价百分之十购票的婴儿无免费行李额。

课后练习和任务

【课后练习】

1. 将下面的英文句子翻译成中文或将中文翻译成英文。

1）Any hand luggage, madam?

2）Could you put on the luggage label，please?

_____

3）Hand-luggage is not weighed.

_____

4）How many pieces of luggage do you want to check in?

_____

5）I mean that one is too long. It's 120 cm long，over the length limitation. I'm sorry，sir. We will charge you for five kilos overweight.

_____

6）我能带这个袋子上飞机吗?

_____

7）需要把我们所有的行李一起称吗?

_____

8）长、宽、高分别不能超过 100 厘米、60 厘米和 40 厘米。

_____

9）恐怕您要付超重行李费了。

_____

10）您付完超重行李费后，带着您的收据回来，我将为您托运行李。

_____

2. 在横线上填写合适的句子，补全对话。

C：Good afternoon. Your ID card and itinerary of e-ticket，please.

P：Here you are.

C：_____ to check in?

P：Yes. Two boxes.

C：Would you put them on the scales? I'm going to weigh them.

P：OK. By the way，can you tell me how many kilos of free baggage each passenger is allowed?

C：_____. What class are you travelling?

P：First class.

C：For a first-class passenger，_____.

P：How many pieces of baggage can I carry onto the plane?

C：Passengers holding first class tickets may carry 2 pieces within the total weight of 5 kg onto the plane.

P：I see. Thank you for your information.

C：You're welcome. Your boarding pass and baggage check，please.

P：Good-bye.

【课后任务】

分小组，根据下面的背景编写对话，要求英文对话不少于 **12** 句，并将对话录制成视频以小组形式上传到学习通平台。

1）You are traveling in Xi'an，but you can't find your suitcase at the airport.

2）The clerk comes and asks you to fill in the "Baggage Transport Accident Record".

3）Two days later，the clerk calls you to claim your luggage.

_____

_____

_____

_____

_____

_____

_____

_____

_____

情景一 Ground Service 地面服务
学习活动四 其他服务（Other Service）

任务描述

假如你是地面问询处工作人员 Anna，现有一外国乘客 Jack 正在你处询问 CA981 次航班延误的相关信息，请你就航班的飞行时间以及延误原因等给予帮助和解答。

任务准备

【任务相关知识准备】请将下面的标识用英语写出来并翻译成中文。

_____

_____

_____

_____

_____

_____

【小组查询资料并完成讨论】 利用网络、手机等媒介，查询资料并讨论：

1）机场地面服务包含哪些内容？

2）机场地面的基本标识有哪些？

3）读下列标识并写出下列标识的中文含义。

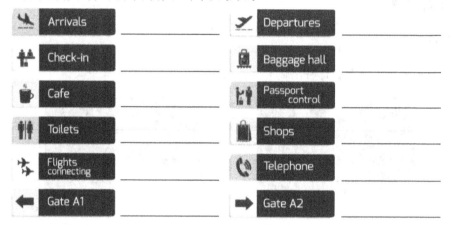

Arrivals _____	Departures _____
Check-in _____	Baggage hall _____
Cafe _____	Passport control _____
Toilets _____	Shops _____
Flights connecting _____	Telephone _____
Gate A1 _____	Gate A2 _____

任务实施

【小组合作探讨完成下列任务】

写一写：分组讨论撰写"Jack 询问航班延误"的对话。

要　求：（1）符合对话逻辑，情节合理；

　　　　　（2）不少于 12 句；

　　　　　（3）语法合乎要求，鼓励创新。

_____

_____

_____

_____

_____

_____

_____

【角色扮演】我的航班怎么了？(Jack 询问航班延误)

演一演：分角色扮演操练（实训场地）。

要　求：(1) 可使用自制的小道具进行展示，达到完成服务工作的目的；

　　　　(2) 在展示过程中注意对话情节合理性和礼貌服务；

　　　　(3) 每组表演时间不超过 5 分钟；

　　　　(4) 2～4 人为小组代表在实训区进行展示；

　　　　(5) 完成小组模拟情境互评打分表。

评价内容	专业能力 70 分			职业素养 30 分						总分	
组别	对话撰写无误，为乘客办理咨询航班信息的手续流程清晰、完整（30 分）			表演中，成员发音清晰，语音语调准确，流程完整，表演流畅（40 分）			对话表演中积极参与，情绪饱满，动作礼仪规范（15 分）			对话表演有一定的创新性（5 分）	对话表演整体效果良好（10 分）	100 分	建议
得分	30	20	10	40	30	20	15	10	5	5	3	1	10	7	4		
一组																	
二组																	
三组																	
四组																	
五组																	

任务拓展和提升

【对话】Dialogue A　Asking for the flight information 询问航班信息

P：Excuse me, I'm going to take flight SH307 to Shanghai, but the board shows it'll be delayed. Do you know what the new departure time is?

R：The flight has been pushed back by about an hour but please keep checking the board just in case.

P：Is it OK for me to go out to have something to eat?

R：You'd better stay at the airport. The flight departure could be announced any minute.

P：What caused the delay?

R：Sorry, I'm not sure.

Dialogue B Flight has been cancelled 航班被取消

Announcer：May I have your attention please? The cancellation of the fight 345 due to technical reasons. All passengers with 4 tickets for this fight please make your way to the Asian Airlines information desk.

A：Which flight has been cancelled?

B：I don't know. But our fight is still on the board. I'll find out.

B：Excuse me，could you tell me which fight has been cancelled?

C：Flight 345 to Los Angeles.

B：Thank you.

> **Key words & expressions**
>
> | board | *n.* 显示牌 | delay | *v.* 延误 |
> | announce | *v.* 广播 | cause | *v.* 导致 |
> | cancellation | *n.* 取消 | due to | 由于 |
> | Los Angeles | 洛杉矶 | departure time | 起飞时间 |
> | in case | 以防万一 | | |

【巩固练习】

1. 分角色朗读上面的对话。

2. 根据对话 A，回答下列问题。

1）Who are the two persons in the dialogue?

_____

2）When will the plane take off?

_____

3）Why did the plane delay?

_____

3. 根据对话 B，回答下列问题。

1）Who are the two persons in the dialogue?

_____

2）What's wrong with the plane?

_____

4. 听老师朗读，写出听到的单词或句子。

1）_____ 2）_____

3）_____ 4）_____

5）_____ 6）_____

7) _____   8) _____

9) _____   10) _____

5. 在横线上填写合适的单词、短语或句子，使对话完整。

P：Do you have a _____ flight from Los Angeles to Xiamen?

A：I'm sorry，we don't have _____ flights from Los Angeles to Xiamen.

P：Are there any other airlines flying _____?

A：I'm afraid no airlines _____

P：How can I get there?

A：Don't worry. I can arrange _____ for you.

P：Do I have enough time to catch _____?

A：According to the flight schedule，you have two hours and a half for _____ and you are sure to make it.

P：I'm interested in taking your flight to Tokyo.

A：_____?

P：October 3.

A：We have two flights to Tokyo every day. _____.

P：I want to take the flight leaving in the afternoon.

A：_____?

P：Book me two tickets.

A：_____.

P：Economy class.

A：_____?

P：No，just one-way trip.

A：_____?

P：My name is Peter Bradly. My wife's name is Susan.

A：_____?

P：You can reach me at 84552244.

6. 请朗读下面的广播稿，注意语音语调。

May I have your attention please? China Eastern Airlines regrets to announce that the departure of flight MU9103 to Beijing will be delayed due to unfavorable weather conditions in Beijing. The new departure time will be announced. Thank you!

任务总结和知识点链接

1. 航班信息主要包含内容：

2. 航班延误的主要原因：

3. 知识点链接——飞机航班号的字母数字表示什么呢？

国内航班一般是四位数字，第一个数字表示始发地区（华北为1，西北为2，南方为"3"，西南为"4"，东南和东部为"5"）；第二个数字表示到达地区（同前）；后两位数字是航空公司航线的编号，其中最后一个数字单数表示去程，双数表示返程。国际航班是三位数。

目前，国内航空公司使用的航班编号采用两位英文字母（或数字加英文字母组合）航空公司代码加上三到四位数航班号的方式，其中国内航班统一为四位数，国际航班为三位数；国内航空公司通过兼并重组，客运航空公司代码如下：CA 代表中国国际航空公司，CZ 代表中国南方航空集团，MU 代表中国东方航空公司，HU 代表海南航空公司，ZH 代表深圳航空公司，FM 代表上海航空公司，MF 代表厦门航空公司，SC 代表山东航空公司，3U 代表四川航空公司；新开办民营航空公司代码如下：EU 代表鹰联航空公司，BK 代表奥凯航空公司，8C 代表东星航空公司，9C 代表春秋航空公司。编码后面的三或四位数，以前国内航班编号还有规律，现在已经打乱了原规则，唯一保持的是去程为奇数，回程为偶数，如中国国际航空公司（国航）基地在北京，国航北京飞广州的航班号是 CA1301 的话，那么从广州飞回北京时航班号就是 CA1302；而南方航空公司（南航）基地在广州，南航广州飞北京的航班号是 CZ3101 的话，回程就变更为 CZ3102。

课后练习和任务

【课后练习】
将下列英文句子翻译成中文。

1) I'd like to reconfirm my flight reservation.

2) That's our regular flight on our record list to Hangzhou.

3) Yes. We have you on our records.

4) Is there any change?

5) At 2:40 p. m., as scheduled.

6）Thank you for calling to reconfirm.

_____

7）You should come to pay and pick up your Itinerary of e-ticket for Air Transportation 2 days before departure.

_____

8）I'd like to confirm my flight to New York.

_____

【课后任务】

分小组，根据下面的背景编写对话，要求英文对话不少于 **12** 句，并将对话录制成视频以小组形式上传到学习通平台。

A woman with a baby is asking the flight information about the Flight CA1379 from Beijing to Shanghai in the information desk. The attendant is helping her to answer the information about the flight（flying time/boarding gate/departure time）.

_____

_____

_____

_____

_____

_____

_____

_____

_____

_____

_____

任务描述

假如你是地面问询处工作人员 Anna，现有一外国乘客 Mike 刚抵达北京机场，他要去往长城酒店，请为其提供相应的帮助，并为其推荐合理的交通工具到达酒店。

任务准备

【小组查询资料并完成讨论】利用网络、手机等媒介，查询资料并讨论完成：

1) 问路常用的句型（写出三个）：

2) 指路的相关英语方位词整理：

3) 地点的英文表达方式

饭店：　　　　　　　　　宾馆：

出租车站：　　　　　　　书店：

火车站：　　　　　　　　购物中心：

银行：　　　　　　　　　地铁站：

任务实施

【小组合作探讨完成下列任务】

写一写：分组讨论撰写"帮助乘客 Mike 在机场打车 去往长城酒店"的对话。

要　求：(1) 符合对话逻辑和情景，不少于 12 句；

　　　　　(2) 语法合乎要求，鼓励创新。

【角色扮演】我要去长城酒店（帮助乘客 Mike 在机场打车去往长城酒店）。

演一演：分角色扮演操练（实训场地）。

要　求：(1) 可使用自制的小道具进行展示，达到完成服务工作的目的；

　　　　(2) 在展示过程中注意情景符合常理和礼貌服务；

　　　　(3) 每组表演时间不超过 5 分钟；

　　　　(4) 2～4 人为小组代表在实训区进行展示；

　　　　(5) 完成小组模拟情境互评打分表。

评价内容	专业能力 70 分			职业素养 30 分									总分	
组别	对话撰写无误，为乘客指路流程清晰、完整（30 分）			表演中，成员发音清晰，语音语调准确，流程完整，表演流畅（40 分）			对话表演中积极参与，情绪饱满，动作礼仪规范（15 分）			对话表演有一定的创新性（5 分）			对话表演整体效果良好（10 分）			100 分	建议
得分	30	20	10	40	30	20	15	10	5	5	3	1	10	7	4		
一组																	
二组																	
三组																	
四组																	
五组																	

任务拓展和提升

【对话】机场问路指路

Dialogue A　How can I get to the city?　怎样去城区？

P＝Passenger；R＝Receptionist

P：Excuse me, how can I get from the airport to the city?

R：You can take a taxi or an airport shuttle.

P：How much is a shuttle ticket?

R：16 Yuan. But you'd better decide which bus line you should take before you get on the bus, sir.

P：Which line should I take if I want to go to Wangjing?

R：Let me check for you. According to the route map, you should take line 6.

Dialogue B　Ask for help　寻求帮助

P：Excuse me, how can I find the taxi pick-up point?

A：You need to get out of the arrival exit, then turn right, and the taxi pick-up point is between Gate 9 and Gate 10.

P：OK，I see. Thanks a lot.

A：My pleasure.

Key words & expressions

shuttle	*n.* 大巴车	ticket	*n.* 票
check	*v.* 检查	according to	根据
pick-up point	出租车上下点		

【巩固练习】

1. 分角色朗读上面的对话。

2. 根据对话，回答下列问题。

1）Who are the two persons in the dialogue A?

_____

2）Where does the passenger want to go?

_____

3）What's wrong with the passenger?

_____

4）How does the receptionist help the passenger?

_____

3. 听老师朗读，写出听到的单词或短语。

1) _____   2) _____
3) _____   4) _____
5) _____   6) _____
7) _____   8) _____
9) _____   10) _____

4. 请根据上下文写出句子，补全对话。

P：I'm interested in taking your flight to Tokyo.

A：_____?

P：October 3.

A：We have two flights to Tokyo every day. _____?

P：I want to take the flight leaving in the afternoon.

A：_____?

P：Book me two tickets.

A：_____.

P：Economy class.

A：_____?

P：No，just one-way trip.

A：_____?

P：My name is Peter Bradly. My wife's name is Susan.

A：_____?

P：You can reach me at 84552244。

5. 请大声朗读下列广播稿。

Ladies and Gentlemen：

May I have your attention，please?

We are paging for Mr. /Ms. _____. Please contact any of our cabin crew.

Thank you!

任务总结和知识点链接

1. 指路常用词汇以及句型。

2. 知识点链接。

1) 候机楼问询为旅客及其他顾客提供诸如航班信息、机场交通、候机楼设施使用等问询服务。

2) 问询服务往往能直接解决旅客在旅行过程中遇到的许多麻烦，或能为旅客解决问题指明方向。

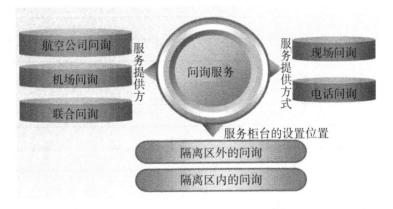

3) 公共广播系统是机场航站楼必备的重要公共宣传媒体设备，是机场管理部门播放航空公司航班信息、特别公告、紧急通知等语言信息的重要手段，是旅客获取信息

的主要途径之一，也是提高旅客服务质量的重要环节。

课后练习和任务

【课后练习】

连线题，请将下列英文短语和对应的中文连起来。

the 1st underground floor   下楼

taxi   在……的左边

sign   远离

on the left of   直走

go downstairs   左转/右转

far away   标志

walk straight   出口

turn left/right   出租车

exit   地下一层

【课后任务】

分小组，根据下面的背景编写对话，要求英文对话不少于 **12** 句，并将对话录制成视频以小组形式上传到学习通平台。

You have just arrived the Beijing International Airport，you want to go to the Yansha Shopping Mall，You ask a clerk in the information.

_____

_____

_____

_____

_____

_____

_____

_____

_____

_____

情景二 Cabin Service 客舱服务

学习活动一 起飞前准备（Pre-flight）

　　假如你是空中乘务员 Alice，请在舱门口处欢迎乘客 Johnson 登机，并引领其至座位。

　　【任务相关知识准备】请将下面框中的单词或词组翻译成中文，并将相对应的英文名称填在图片箭头相应的横线上。

first name _____	family name _____	flight number _____
boarding time _____	gate number _____	seat number _____
date _____	airline _____	

　　【小组查询资料并完成讨论】利用网络、手机等媒介，查询资料并讨论：

　　一般登机牌的构成有：_____

_____

_____

任务实施

【梳理常用语句】

1) Good morning/afternoon/evening，sir/madam.

2) How do you do?

3) How are you?

4) Nice to meet you.

5) May I see your boarding pass，please?

6) See you later.

7) Have a nice trip.

8) Wish you a happy stay in China.

9) Welcoming on board.

10) Hello there，how are you today?

11) Could I please see your boarding pass?

12) Would you mind just taking this seat until I have checked the passenger list?

【小组合作探讨完成下列任务】

写一写：分组讨论撰写"欢迎乘客 Johnson 登机"的对话。

要　求：(1) 符合欢迎乘客登机的流程；

　　　　(2) 内容合情合理，不少于 12 句对话；

　　　　(3) 语法合乎要求，鼓励创新。

_____

_____

_____

_____

_____

_____

_____

_____

_____

【角色扮演】欢迎乘客 Johnson 登机。

演一演：分组角色扮演操练（实训场地）。

要　求：(1) 可使用自制的小道具进行展示，达到完成服务工作的目的；

　　　　(2) 在展示过程中注意办理欢迎乘客登机的流程和礼貌服务；

　　　　(3) 每组表演时间不超过 5 分钟；

（4）2～4人为小组代表在实训区进行展示；

（5）完成小组模拟情境互评打分表。

评价内容	专业能力 70 分						职业素养 30 分								总分	建议
组别	对话撰写无误，欢迎乘客 Johnson 登机手续流程清晰、完整（30分）			表演中，成员发音清晰，语音语调准确，流程完整，表演流畅（40分）			对话表演中积极参与，情绪饱满，动作礼仪规范（15分）			对话表演有一定的创新性（5分）			对话表演整体效果良好（10分）			100分	
得分	30	20	10	40	30	20	15	10	5	5	3	1	10	7	4		
一组																	
二组																	
三组																	
四组																	
五组																	

任务拓展和提升

【对话】**Dialogue A   Welcome aboard   欢迎登机**

A＝attendant；P＝passenger

A：Good morning, sir. Welcome aboard!

P：Good morning, where is my seat?

A：May I see your boarding pass, please?

P：Yes. Here it is.

A：35D. Walk straight ahead to the rear. The stewardess there will show you where your seat is.

Dialogue B   Direct to the Assigned Seat   引导座位

F＝flight attendant；P＝passenger

F：Good morning/afternoon evening. Welcome aboard!

P：Good morning/afternoon/evening. Could you show me where my seat is?

F：May I see your boarding pass, please?

P：Here you are.

F：Your seat number is 54A, in the rear of the cabin and turn right. The seat number is on the edge of the rack.

P：Thank you. And when will we take off? And the time to land?

F：We will take off at 9：00 a. m. and land at 11：30 a. m..

P：Which terminal will we arrive at?

F：We will arrive at Terminal 1.

P：Thanks.

F：You are welcome.

Key words & expressions

straight	*adj.* 直的	stewardess	*n.* 女乘务员
rear	*adj.* 后面的	terminal	*n.* 航空站；航空终点站
land	*v.* 降落，着陆	take off	起飞
boarding pass	登机牌	Welcome aboard	欢迎登机

【巩固练习】

1. 分角色朗读上面的对话。

2. 根据对话回答下列问题。

1）Who are the two persons in the dialogue?

_____

2）Where is the passenger's seat?

_____

3. 听老师朗读，写下所听到的单词或句子。

1）_____   2）_____

3）_____   4）_____

5）_____   6）_____

7）_____   8）_____

9）_____   10）_____

4. 在横线上填写合适的单词、短语和句子，使对话通顺。

第一篇

Mr. John（J）is boarding the plane and a flight attendant（A）is welcoming him.

A：Good morning, sir. _____ .

J：It's nice to be in your plane.

A：May I see your _____ , please?

J：Sure, _____ .

A：Please _____ to the middle. Your seat is on your right by the window.

第二篇

A：Your seat number is 22D. It's at the back of the _____ , the 2nd _____ from the last. It's the only by the aisle.

P₁：_____ . But would it be possible for me to have a window seat?

A：It's OK if the seat is _____ , sir.

（A passenger is blocking the way in the cabin. A stewardess comes to him. ）

A：Excuse me，sir. Would you _____ stepping aside and _____ the other passengers pass ?

P₂：Oh，sorry.

第三篇

P₁：Miss，the man in 16E is my friend. Can I _____ with him?

A：Just a _____ . I will _____ to talk with that lady in 16D.

（To P₂）：_____ me，madame. That gentleman over _____ is a friend of this young man beside you. They'd like to sit _____ . Would you mind _____ with him?

P₂：No ，I _____ .

A：Thank you，and _____ for the trouble.

任务总结和知识点链接

1. 乘客登机需要准备的材料：

2. 登机牌的构成：

3. 知识点链接——欢迎乘客登机的注意事项以及流程：

1）欢迎旅客登机时，乘务员应着装整齐、站姿端正、面带微笑、热情有礼、迎接问候，并同时引导客人入座，协助放好物品，之后根据乘客要求提供服务，如报纸杂志、热毛巾等。

此任务有两个重要的环节：①迎宾。以饱满的热情，迎接每位乘客的登机。迎宾礼仪是空乘直接服务于乘客的第一步，给乘客留下的心理感受将影响其对公司服务的评价，必须给予高度重视。②问候。"欢迎您登机"，送出第一声问候，代表机组成员对乘客的真诚与欢迎。

2）引导乘客入座时要注意：①乘务员应均匀分布在各自岗位区域，站姿端正、优雅，面带微笑，目光柔和亲切，向每一位乘客微笑致意，主动问候，引导入座。②及时疏通过道，以免影响乘客正常登机。③乘务员在引导乘客入座时，应主动帮助安排乘客将行李摆放到安全、合适的位置。

3）乘客登机相关知识点。

（1）乘客登机：登机时间一般在起飞前的 20～30 分钟；起飞前 5 分钟（无托运行李的旅客）/起飞前 10 分钟（有托运行李的旅客）不能到达登机口的旅客，将不能登机。听到登机广播后，在登机口会有服务人员检查登机牌，乘客持检查完的登机牌登机。

（2）寻找机上位置。登机牌上标明有机上座位的位置，如 5D、11C，数字代表第几排，字母代表每排座位编号，飞机上的座位号标在放行李的舱壁（座位上方）。

（3）飞机上的座位是如何分布的？

一般情况下，中小型客机为每排 4～6 座，中间为通道；大型客机或宽体客机为双通道，每排 7～9 座（普通舱）。座位的编号或为阿拉伯数字，或为英文字母，但都是从第一排开始排至最后。每排座位从左至右编为 A、B、C、D、E、F 等。如果登机牌号为 30A，所乘飞机型号为波音 737，则您的座位是 30 排左边靠舷窗 "A" 的位置。

787

布局图

■ 豪华头等舱　■ 头等舱　□ 公务舱　■ 高端经济　□ 经济舱　△ 出口

■ 厨房　■ 清洗室　■ 衣帽间　● 婴儿摇篮挂点位置　▲ 紧急出口

课后练习和任务

【课后练习】

1. 将下列单词或词组补充完整并熟记。

1. 直走：Go _____. 穿过：_____ the cabin. 向右转：_____ right.
2. 在前：in the _____. /中：_____ /后：_____ of the cabin（舱）
3. 靠窗座位：_____. 靠走道座位：_____.
4. 起飞：_____. 落地_____.

2. 请将右列的单词或词组填在左边对应的句子中，使句子完整。

1) Would you please take the _____ till after taking off? I will try to _____ it later.	A. vacant seat
	B. arrange
2) I'm sorry there is no _____ on board.	C. in the rear
3) There are some vacant seat _____, Would you like to sit there?	D. assigned seat

3. 请朗读下面的句子并熟记。

Polite Requests： Can I see your boarding pass?	Please/Can I check the seating arrangements?
Can I look at your seat number，please?	Please/Would you sit here for the moment?
Or，more politely： Could I see your boarding pass，please?	Would you follow me，please?
	Would you please turn off your mobile phone?
Could I check your seat number，please?	Would you mind just taking this seat until I have checked the passenger list?

Euphemism（委婉语）

Could/Would/May/Might/Shall... ?

Why don't you.../Do you mind＋V-ing... ?

What about/How about＋V-ing... ?

I'm sorry...

I wonder/hope，I wondered.

I was wondering/I was hoping...

... ，but...

4. 不改变句意，用括号里的句型改写下列句子。

1) I want to see your boarding pass. （Could I...）

2) Put your bag in the overhead locker. （Would you...）

_____

3) What's your seat number? （May I...）

_____

4) Don't use the lavatory. （I'm afraid...）

_____

5. 请在横线上填写相对应的英文单词或短语，使句子完整。

1) By the way, if there are some valuables in your baggage, you may _____ them out（取出）and _____ （放置）them in the seat _____ （口袋）in _____ （前方的）of you.

2) _____ （保管好）your valuables，please.

3) as _____ as _____ （尽快）.

4) _____ （托运）行李：_____ （提取）.

5) I will get them to you _____ （马上）.

6) 用三种以上的说法表达"马上、立刻"。

① _____ ② _____

③ _____ ④ _____

【课后任务】

分小组，将下面的情景编成对话并录制成视频，以小组形式上传到学习通平台。

A woman with a baby is boarding，the attendant is helping her to find the seat.

_____

_____

_____

_____

_____

_____

_____

任务描述

　　飞机舱门已经关闭，等待塔台信号即将起飞。假如你是空中乘务员 Alice，请你进行客舱内安全检查，提醒乘客 A 关闭手机，告知乘客 B 系好安全带，准备起飞，即将开始一段美好的航程。

任务准备

　　【任务相关知识准备】 你能总结出介绍紧急出口的关键点吗？请试着按照提示用英语写出来。

Key1	说明紧急出口在何时使用以及使用方法
Key2	介绍旅客的座位靠配和小桌板的使用方法
Key3	紧急出口行李要求
Key4	询问旅客是否听懂以及是否愿意坐在这里
Key5	如果愿意，请旅客阅读《安全须知卡》并且表示感谢；如果不愿意，帮助旅客调换座位

Key 1 _____
Key 2 _____
Key 3 _____
Key 4 _____
Key 5 _____

　　【小组查询资料并完成讨论】 利用网络、手机等媒介，查询资料并讨论：

(1) 飞机即将起飞时，为什么要求乘客关闭手机以及电子设备？

(2) 飞机即将起飞时的安全检查工作有哪些？

任务实施

【小组合作探讨完成下列任务】

写一写：分组讨论撰写"进行飞机起飞前安全检查"的对话。

要　求：（1）符合飞机起飞前安全检查的流程；

　　　　　（2）要求情节设计合理，不少于 12 句；

　　　　　（3）语法合乎要求，鼓励创新。

_____

_____

_____

_____

_____

_____

_____

【角色扮演】即将开启一段愉快的航程。（进行飞机起飞前安全检查）

演一演：分组角色扮演操练（实训场地）。

要　求：（1）可使用自制的小道具进行展示，达到完成服务工作的目的；

　　　　　（2）在展示过程中注意安全检查的流程和礼貌服务；

　　　　　（3）每组表演时间不超过 5 分钟；

　　　　　（4）2～4 人为小组代表在实训区进行展示；

　　　　　（5）完成小组模拟情境互评打分表。

评价内容	专业能力 70 分						职业素养 30 分									总分	
组别	对话撰写无误，进行飞机起飞前安全检查流程清晰完整（30分）			表演中，成员发音清晰，语音语调准确，流程完整，表演流畅（40分）			对话表演中积极参与，情绪饱满，动作礼仪规范（15分）			对话表演有一定的创新性（5分）			对话表演中整体效果良好（10分）			100分	建议
得分	30	20	10	40	30	20	15	10	5	5	3	1	10	7	4		
一组																	
二组																	
三组																	
四组																	
五组																	

任务拓展和提升

【对话】Dialogue A　Emergency Exit Seats Introduction 紧急出口座位须知

F：And you are sitting next to the emergency. Please allow me to introduce some for you.

P：OK.

F：This red handle is for emergency using，please don't press it normally. Mean while，please prevent other passengers from touching it. You are supposed to assist us in this door in case of emergency. Please keep the emergency exits clear of baggage. Your seat back can not be adjusted. Your tray table and screen are in your armrest，please stow them after using. Here is the Safety Instruction Card，please read it before take-off. If you would like to change your seat for a non-exit seat，please tell me，I will arrange it for you got it，sir?

P：Yes.

F：Are you willing to sit here?

P₁：No problem.

F：Thank you for your time. And keep your seat belt fastened，please.

P₂：Oh，I don't want to sit here.

F：It doesn't matter. I will try to arrange another seat for you.

Dialogue B　Security check　客舱安全检查

The chief attendant：（Announcement）Good morning，ladies and gentlemen. Welcome you aboard Airline to Shenzhen. Please make sure that your seat belt is fastened，your seat is upright，and your tray table is closed. Your cabin baggage should be in the overhead compartment or under the seat in front of you. No smoking will be permitted on this flight.

A：Excuse me，madam . Would you please return your seat back to the upright position?

P：Oh，I'm sorry，but how can I return it back?

A：Just press the button on your armrest.

…

P：Is that right?

A：OK，Please also check that your seat belt is fastened.

P：Thank you!

A：It's my pleasure，have a nice trip!

Dialogue C　Turn off Electronic Devices　关闭电子通信设备

FA：Excuse me，sir. Please turn off your mobile phone. We are going to take off soon.

PAX：Why? My mobile phone is in flight-mode.

FA：Because flight-mode is also against CAAC regulation. The mobile phone must be turned off during the whole flight.

PAX：Oh，I see. I will use my lap-top to do some work.

FA：Sorry，sir. Electronic devices can only be used once we reach our altitude. Thank you for your understanding.

PAX：Fine. Thanks for reminding me.

Key words & expressions

emergency	*n.* 紧急出口	allow	*v.* 允许
necessary	*adj.* 有必要的	information	*n.* 信息
press	*v.* 按下	adjust	*v.* 调节
armrest	*n.* 扶手	stow	*v.* 把……收好
fasten	*v.* 系好	check	*v.* 检查
chief	*adj.* 主要的	announcement	*n.* 广播
airline	*n.* 航班	tray	*n.* 托盘
permit	*v.* 允许	device	*n.* 设备
altitude	*n.* 海拔高度		
make sure	确认	seat belt	安全带
Have a nice trip!	祝您旅途愉快！	mobile phone	手机

【巩固练习】

1. 分角色朗读上面的对话。

2. 根据对话 A，回答下列问题。

1）Who are the two persons in Dialogues A?

_____

2）Where did the passenger sit?

_____

3）Who would like to sit next to the emergency?

_____

3. 根据对话 B，回答下列问题。

1）Who are the two persons in Dialogues B?

2）Where did the flight going?

4. 根据对话 C，回答下列问题。

1）Who are the two persons in Dialogues C?

2）Why can't the Electronic devices be used?

5. 从右列所给的选项中选出与左列动词符合的内容并将句子翻译成中文。

1）greet   a. everyone to switch off mobile phones

2）check   b. the exit row is clear

3）make sure that   c. that all seat belts are fastened

4）close   d. passengers

5）show   e. the overhead lockers

6）tell   f. the safety instruction card

6. 请熟读下列句子并翻译成中文。

This red handle is for emergency using，please don't press it normally. Meanwhile，please prevent other passengers from touching it. You are supposed to assist us in this door in case of emergency. Please keep the emergency exits clear of baggage. Your seat back can not be adjusted. Your tray table and screen are in your armrest，please stow them after using. Here is the Safety Instruction Card，please read it before take-off. If you would like to change your seat for a non-exit seat，please tell me，I will arrange it for you.

7. 听老师朗读，写下所听到的单词或句子。

1) _____   2) _____

3) _____   4) _____

5) _____   6) _____

7) _____   8) _____

9) _____   10) _____

8. 请朗读下面的广播稿。

Good morning (afternoon/evening)，Ladies and Gentlemen：

Welcome aboard _____ (Airlines).

I am _____, the purser for this flight. During the flight，all of my colleagues will be happy to be of service to you. Thank you.

Ladies and Gentlemen. We will be taking off in a few minutes. Please be seated and fasten your seat belt (Please stow your foot rest). Your seat back and table should be returned to the upright position.

All _____ (Airlines) flight are non-smoking to comply with government regulation. Please refrain from smoking during the flight.

Thank you for your cooperation，and we wish you a pleasant journey！

任务总结和知识点链接

1. 在起飞前，乘务员要进行安全确认的工作内容。

2. 知识点链接——起飞前的民航安全管理规定。

飞机在滑行时，乘务长要进行广播或派其他人广播，进行安全说明。

- 紧急设备的位置和使用
- 出口位置和使用
- 紧急程序
- 可以帮助实施紧急程序的旅客

在飞机起飞前，乘务员必须做到：

- 确保卫生间无人占用
- 调暗客舱灯光
- 关闭厨房电源

- 旅客座椅处无饮料和餐具
- 在值勤的位置上坐好，系好安全带和肩带

如果客舱还未做好起飞准备，乘务长应通知机长。

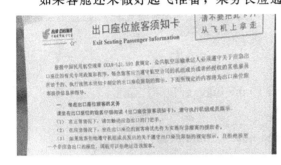

知识点链接——乘坐飞机时的安全注意事项有哪些？

（1）登机后起飞前，空服人员会进行安全提示，需要留意一下他们的提示；

（2）在起飞至降落的全过程中，需要按照空服人员的提示关闭移动电子设备；

（3）在起降及遇到气流产生颠簸时，要系好安全带；在飞行过程中，无特殊情况也建议系好安全带；

（4）留意最近的紧急出口的位置，在出现紧急情况时，第一时间有秩序撤离；

（5）逃生时不要携带行李！

知识点链接——紧急情况使用语言。

（1）Remove sharp objects，all the shoes/high-heeled shoes！Put them in your carry-on baggage. Fasten your seat belt. Stow your tray table. Bring your seat back to the Position. Open the windows-shade.

取下尖锐物品，脱下（水上：鞋子/陆地：高跟鞋）放在行李中。系好安全带，收起小桌板，调直座椅靠背，打开遮光板。

（2）Take out your life vest under your seat！Slip it over your head！

从座位下取出救生衣！经头部穿好！

（3）Fasten the Buckles！Don't inflate in the cabin！

带子扣好，系紧！客舱内不要充气！

（4）Attention please：Evacuate from this exit. If this exit doesn't work，go to that one.

这排的旅客注意了：飞机完全停稳后，从这个出口撤离，如果这个出口不能使用，从那个出口撤离！

（5）Which way do you evacuate？If this exit doesn't work，which exit？Right（wrong）！

你从哪个出口撤离？如果这个不能用，从哪个呢？正确（错误）！

（6）Fasten seat belt，take safety position.

系好安全带，做好防冲撞姿势。

（7）Follow me!

跟我做！

（8）Take safety position! I can assist you!

保持姿势，我来检查！

（9）Right，relax!

正确，放松！

（10）Open seat belt! Don't move. Don't take luggage!

解开安全带！不要动！别拿行李！

（11）Stay down! Cover your mouth and nose! Come here!

低下身！捂住口鼻！到这边来！

（12）Come here! One by one/two at a time! Jump! Slide!

到这边来！一个接一个/（双通道：两个一起）！跳！滑！

（13）Come here! Inflate life vest! Jump into the water! Board the raft!

到这边来！救生衣充气！下水！上船！

（14）No exit，go that way.

此门不通，到那边去.

（15）Anybody here? Answer me!

还有人吗？听到请回答！

（16）Passenger in this row，attention please：evacuate from this exit，if this exit doesn't work，go to that one.

这排的旅客注意了：飞机完全停稳后，从这个出口撤离，如果这个出口不能使用，从那个出口撤离！

课后练习和任务

【课后练习】
请将下列单词连成通顺的句子并翻译成中文。

1）seat belt please your sir fasten

_____

2）upright the your in please seat position put

_____

3）seat belt just I your can check

_____

4）mind you off would computer switching your

_____

5）phone off please switch your

_____

6）table sir up your you could put

_____

7）bag mind the putting in overhead would your locker you

_____

【课后任务】

分小组，根据下面的背景编写对话，要求英文对话不少于 12 句，并将对语录制成视频以小组形式上传到学习通平台。

A man want to smoke in the aisle，the attendants stopping the passenger.

_____

_____

_____

_____

_____

_____

_____

_____

检查安全与手机子任务结束

情景二　Cabin Service 客舱服务
学习活动二　飞行中服务（Cabin Service）

　　飞机已经进入平飞阶段，并且到了提供午餐的时间，假如你是空中乘务员 Alice，请为乘客提供餐饮服务。

【任务相关知识准备】请将下列图片对应的英文填在横线上。

【小组查询资料并完成讨论】

1. 机上餐食的种类以及特殊餐食的种类：

2. 机上服务的基本程序和要求：

任务实施

【梳理餐食词汇，并熟读记忆】

breakfast	早餐	lunch	午餐
dinner	正餐	menu	菜单
Chinese food	中餐	western food	西餐
steamed	蒸的	fried	油炸的

COOKING　烹调

roast	烤的	taste	味道
fresh	鲜的	sour	酸的
bitter	苦的	crisp	脆的
hard	老的	delicious/tasty	美味的
light	清单的	tableware	常用餐具
bowl	碗	dish	碟子
spoon	勺	cup	杯子
knife	餐刀	fork	叉子
toothpick	牙签		

FOOD　主食及其他

porridge/congee	粥	scrambled eggs	炒蛋
omelette	煎蛋卷	frittata	肉馅煎蛋饼
dimsum	点心	noodles	面条
fried noodles	炒面	fried rice	炒饭
dumpling	饺子	bread	面包

【小组合作探讨完成下列任务】

写一写：分组讨论撰写"为乘客 Mike 提供餐食"的对话。

要　求：(1) 符合为乘客提供餐食的流程；

（2）要求情节设计符合逻辑，不少于 12 句；

（3）语法合乎要求，鼓励创新。

【角色扮演】 开饭了！（为乘客 Mike 提供餐食）

演一演：分组角色扮演操练（实训场地）。

要　求：(1) 可使用自制的小道具进行展示，达到完成服务工作的目的；

　　　　　(2) 在展示过程中注意为乘客提供餐食的流程和礼貌服务；

　　　　　(3) 每组表演时间不超过 5 分钟；

　　　　　(4) 2～4 人为小组代表在实训区进行展示；

　　　　　(5) 完成小组模拟情境互评打分表。

评价内容	专业能力 70 分			职业素养 30 分			总分	建议
组别	对话撰写无误，欢迎乘客 Mike 提供餐食流程清晰完整（30 分）	表演中，成员发音清晰，语音语调准确，流程完整，表演流畅（40 分）	对话表演中积极参与，情绪饱满，动作礼仪规范（15 分）	对话表演有一定的创新性（5 分）	对话表演中整体效果良好（10 分）	100 分		建议
得分	30　20　10	40　30　20	15　10　5	5　3　1	10　7　4			
一组								
二组								
三组								
四组								
五组								

任务拓展和提升

【对话】 Dialogue A　Serve Drinks　提供饮料

FA：What would you want to have, madam? We have prepared soda, juice, coffee, tea and mineral water.

PAX：I'd like some juice.

FA：Which flavor do you like? We have orange and coconut.

PAX：Coconut，please.

FA：With ice?

PAX：No, thanks.

FA：Here you are.

Dialogue B　Ask about the second Meal　询问餐食

PAX：Excuse me, miss, Will you serve a meal later?

FA：Yes，sir. We serve two meals during the flight. Next one is two hours later. And we have prepared some snacks and drinks at the bar. Please feel free to enjoy then.

PAX：Great!

FA：And you can also get toothbrushes and eye shaded there. Have a good sleep!

PAX：Thanks. And please don't wake me up then. I want to sleep a little longer.

FA：OK! You may tell us if you need your breakfast by then.

PAX：Thanks a lot.

FA：You're most welcome.

Key words & expressions

prepare	v. 准备	toothbrush	n. 牙刷
snack	n. 小吃	mineral water	矿泉水

【巩固练习】

1. 分角色朗读上面的对话。

2. 根据对话回答下列问题。

1）What drink did the passenger have in the dialogue A?

_____

2）What flavor did the passenger have in dialogue A?

_____

3）What's the request for the passenger in dialogue B?

_____

4）Did the air serve the second meal? When?

_____

3. 听老师朗读，写下所听到的单词或句子。

1）_____　　2）_____

3）_____　　4）_____

5）_____　　6）_____

7）_____　　8）_____

9）_____　　10）_____

4. 在横线上填写合适的单词、短语或句子，使对话通顺。

A：What would you _____, sir? Coffee, tea or fruit juice?

P：A cup of hot tea, please.

A：Green tea or black tea?

P：I _____ jasmine tea.

A：I'm awfully _____ but we don't have that on _____. Is there anything

_____ you'd like?

P：OK，I'll have black tea.

A：_____ .

P：Many thanks.

5. 请熟读下列广播稿。

Before Dinner

Ladies and Gentlemen：

May I have your attention，please! We'll be serving breakfast（lunch/dinner）soon. Welcome to make your choice. If you have special diet requirement，please tell the flight attendant.

Thank you!

Drink Service

Ladies and Gentlemen：

We shall be serving tea，coffee，cola，sprite，orange juice and mineral water in a moment . You are welcome to take your choice.

Thank you!

任 务 总 结 和 知 识 点 链 接

1. 提供餐食的基本流程：

2. 知识点链接。

飞机餐是民航飞机在航程中供应给乘客的餐饮。飞机餐菜式由航空公司定，一般由指定供应航机饮食的机构供应，在机场附近制作，并直接运送至航机上，在航程中途且飞机稳定时由空中服务员放在手推车上分发予乘客。不同等级客位的飞机餐，在菜式、分量及成本各方面都有分别。头等及商务客位的飞机餐，食物及进餐程序皆尽量模仿高级餐厅，虽然如此，但与真正的餐厅始终有别。而经济客位的飞机餐，则与快餐较为相似，以分发效率、储存体积及成本等为主要考虑因素。

在中长途航线中，可能会有超过一顿飞机餐，由于时差关系，通常每餐会隔5～6小时供应，就算当时的所在地并非用餐时间，亦可能在两餐之间提供如杯面等食物。在短程航线中，如两小时以内的国内或国际航线，则只供应一些小食、点心或三明治及饮品。一些极短程的航线则更只有饮品提供。飞机餐的费用一般已包含在机票价格里。除餐食外，航空公司亦会提供软性饮料，如汽水、果汁、茶等饮品，以及酒精类饮品，如餐酒。部分航空公司的酒精类饮品需要额外收费。

一般经济客位的飞机餐包括一包如花生之类的餐前小食，正餐以餐盘盛载，由空中服务员一次分发予乘客。正餐一般包括头盘或沙律，以肉类（如牛或鸡肉）作主菜，伴以饭、面或意大利粉，亦有蛋糕等甜品，部分航空公司更会以冰激凌作甜品。餐后会供应咖啡或茶。

头等及商务客位的飞机餐，在菜式上比经济客位较为多元化，餐前小食亦是花生之类，头盘及主菜像餐厅般以碟盛载，由空中服务员依次端上。头盘除西式外，亦可能有寿司等，部分航空公司有餐汤供应，主菜亦与餐厅类似，如牛柳，亦有较精致的甜品，餐后亦供应咖啡或茶。

特别餐：如果乘客有体质或宗教需要，一般可于出发前 24 小时通知航空公司准备。特别飞机餐包括儿童餐、供糖尿病等慢性病乘客享用的低盐或低糖餐、素食餐、特别宗教要求例如犹太教或回教餐等。

【课后练习和任务】

【课后练习】

1. 写出下列饮料的英文。

(1) _____ （芒果汁/椰汁）

(2) _____ （凉茶）

(3) _____ （矿泉水/纯净水）

(4) _____ （雪碧/七喜）

(5) _____ （百事可乐/可口可乐）

(6) _____ （纸杯）

(7) _____ （咖啡/茶）

2. 学习下列情景常用语并熟记。

1）介绍

We have prepared tomato and egg noodles, and beef rice today.

We have tomato and egg noodles, and beef rice today.

2）询问

Which one do you like/care for?

Which one do you prefer?

What would you like to drink?

Would you like some...?

Care for some coffee?

3）提供服务

This is your...

Be careful. It's very hot.

Here is your...

Enjoy your meal.

Here you are.

4）道歉

Sorry，we are out of it.

Sorry，we don't have any special meal on board.

5）弥补

Let me check if there are some avaiable in other _____ ...

But don't worry, sir. We have some fruits and desserts，would you like some instead?

6）您还可以...

If you need the special meal，you may reserve it 24 hours before departure next time.

7）收桌

Did you enjoy your meal，sir?

I hope you enjoyed your dinner.

May I clean up your tray table，sir?

Let me clean up your tray table.

3. 将下列答语与问题匹配连线。

1）I'd like the beef, please.

a. Certainly, Earl Grey or English Breakfast?

2）Have you got any paracetamol?

b. I do apologize, we've been so busy.

3）I'd like some tea，please.

c. I am sorry, but we've run out of beef.

4）I rang the call button several times.

d. I'm really sorry, we've only got beef left.

5）The sound of the film still doesn't work.

e. I can only apologize. I'll change it.

6）This meal's sold.

f. Sorry about that，try pushing this button.

7）I'd like the fish，please.

g. Yes，but I am afraid I can't give it to you without the doctor.

【课后任务】

分小组，根据下面的背景编写对话，要求英文对话不少于 **12** 句，并将对语录制成视频以小组形式上传到学习通平台。

Please serve the meal for a vegetarian passenger. (Use the sentences given upright)

_____

_____

_____

_____

_____

任 务 描 述

　　飞机已经进入平飞阶段，假如你是空中乘务员 Alice，请你为乘客 Mike 提供报纸和杂志服务。

任 务 准 备

【小组查询资料并完成讨论】

1. 机上可提供的娱乐休闲和服务设施有哪些？

2. 为乘客提供杂志书刊的服务礼节有哪些？

任 务 实 施

【梳理词汇和句型】 请熟记机上娱乐设施相关词汇。

newspaper	报纸
magazine	杂志
Global Times	《环球时报》
China Daily	《中国日报》
device	装置
entertainment	娱乐
pleasure	快乐
touch	触摸
screen	屏幕
operate	操作
movie	电影
choose	挑选
cruise	巡航
headset	耳机

【小组合作探讨完成下列任务】

写一写：分组讨论撰写"为乘客 Mike 提供报纸"的对话。

要　　求：(1) 符合为乘客提供报纸服务的流程；

　　　　　　(2) 要求情节设计符合逻辑，不少于12句；

　　　　　　(3) 语法合乎要求，鼓励创新。

_____

_____

_____

_____

_____

_____

_____

_____

_____

_____

【角色扮演】帮我来一份 China Daily。（为乘客 Mike 提供报纸）

演一演：分组角色扮演操练（实训场地）。

要　　求：(1) 可使用自制的小道具进行展示，达到完成服务工作的目的；

　　　　　　(2) 在展示过程中注意为乘客提供报纸服务的流程和礼貌服务；

　　　　　　(3) 每组表演时间不超过5分钟；

　　　　　　(4) 2～4人为小组代表在实训区进行展示；

　　　　　　(5) 完成小组模拟情境互评打分表。

评价内容	专业能力 70 分						职业素养 30 分								总分	
组别	对话撰写无误，为乘客提供报纸服务手续流程清晰完整（30分）			表演中，成员发音清晰，语音语调准确，流程完整，表演流畅（40分）			对话表演中积极参与，情绪饱满，动作礼仪规范（15分）			对话表演有一定的创新性（5分）			对话表演中整体效果良好（10分）			100分	建议
得分	30	20	10	40	30	20	15	10	5	5	3	1	10	7	4		
一组																	
二组																	
三组																	
四组																	
五组																	

任务拓展和提升

【对话】Dialogue A　Entertainment Device　机上娱乐

PAX：Excuse me，Miss Rose. Can you help me with this device?

FA：Sure. This is our on-board entertainment device Mini-Pad，you can use it to relax and please during your flight.

PAX：So，how can I use it?

FA：As you can see，it has a touch screen，which makes is very easy to operate. There are quite a lot of movies，music and games in it for you to choose. But it can only be used during the cruising period. When you finish using it，please put it in the sit poket in front of you.

PAX：I got it. Thanks a lot.

FA：It's my pleasure，sir. Enjoy your flight.

（after a while）

PAX：Excuse me，my headsets doesn't seem to work. Can you give me hand?

FA：Sorry，sir. They are probably broken. Let me get you a new pair. Please wait a moment.

PAX：Thank you.

FA：You are welcome，sir.

Dialogue B　Serve Magazines　提供报纸

FA：We are terribly sorry，madam. The economy class is full today. We apologize for the inconvenience. Here are some English magazines and newspapers. Anyway，enjoy your flight.

PAX：Thank you any way.

Key words & expressions			
cruising	*v.* 巡航	period	*n.* 阶段
headset	*n.* 耳机	operate	*v.* 操控
screen	*n.* 屏幕	inconvenience	*adj.* 不便的
magazine	*n.* 杂志	apologize for	为……道歉
entertainment device	娱乐设备	economy class	经济舱

【巩固练习】

1. 分角色朗读上面的对话。

2. 根据对话 A，回答下列问题。

1）Who are the two persons in Dialogue A?

_____

2）What kind of device does the flight have?

_____

3）What's wrong with the device?

_____

4）How does the attendant deal with the headsets?

_____

3. 根据对话 B，回答下列问题。

1）Who are the two persons in Dialogue B?

_____

2）Where did the flight going?

_____

4. 请将下列句子翻译成中文。

1）Did you call，sir?

_____

2）What's the problem?

_____

3）What's the matter?

_____

4）How can I help（you）?

_____

5）Can you show me how it works? (Passenger question)

_____

6）Of course. /Certainly. （Flight attendant answer）

_____

7）This is how it works.

_____

8）Is that OK/All right with you? （Checking understanding）

_____

5. 听老师朗读，写下听到的单词或句子。

1) _____ 2) _____

3) _____ 4) _____

5) _____ 6) _____

7) _____ 8) _____

9) _____ 10) _____

6. 朗读下面的广播稿，注意语音语调。

Ladies and Gentlemen：

We are sorry that the video system is not available on this flight，but you can have a wide selection of audio programs. We sincerely apologize for the inconvenience.

Thank you for your understanding.

任 务 总 结

1. 为乘客提供杂志书刊的流程服务礼节：

2. 知识点链接——熟记机上娱乐设施相关词汇。

choose	挑选
cruise	巡航
headset	耳机
give me a hand	帮忙
probably	很可能地
broken	坏掉了
restart	重启
confirm	确认
normally	正常地
belonging	所有物；行李
recommendation	推荐
program	节目

Facilities in a Plane 机上设备

call button	呼叫铃
air-flow knob	通风器
reading light	阅读灯
remote controller	手柄
channel selector	频道选择
volume control	音量控制
VCC（Video Control Center）	视频控制中心
VCS（Video Control System）	视频控制系统

知识点链接：飞机上可以使用电子设备或者手机吗？

iPad、Mp4、照相机都可以使用的，但是在飞机起飞和降落时不能使用，所有电子设备都要关闭，进入平飞状态的时候，有些航空公司的航班可以使用这些电子设备。

手机是飞行全程都不能使用的，虽然现在手机都有了飞行模式，但是手机还是会接收无线电信号，这样就会影响航班飞行，而且现在很多航空公司已经禁止使用手机的飞行模式，所以为了自己和他人的安全，请不要使用手机。

课后练习和任务

【课后练习】

1. 请在下列横线上填上合适的英文，将句子补充完整。

1. We have some English _____（杂志）and _____（报纸）.

2. _____（一份）*Global Times* please.

3. It can only be used during the _____（平飞阶段）.

4. The headsets are probably _____（坏了）. Let me get you a new one.

5. I'll be back and _____（确认）for you soon.

6. The entertainment system can _____（正常的）be used.

2. 请用括号里所给的句型改写句子。

1）I want to see your boarding pass. （Could I...）

_____

2）But your bag in the overhead locker. （Would you...）

_____

3）What's your seat number? （May I...）

_____

4）Don't use the lavatory. （I'm afraid...）

_____

【课后任务】

分小组，根据下面的背景编写对话，要求英文对话不少于 **12** 句，并将对语录制成视频以小组形式上传到学习通平台。

A man want to see a movie in a plane, the attendants helping the passenger to operate the device.

_____

_____

_____

_____

_____

_____

_____

_____

任务描述

飞机已经进入平飞阶段，假如你是空中乘务员 Alice，请为乘客进行免税品销售服务。

任务准备

【小组查询资料并完成讨论】

1. 机上可进行免税品销售的物品有哪些？

2. 机上免税品销售的流程是什么？

任务实施

【梳理词汇和句型】请写出下列词汇的英语表达。

香水　　　　　　　　　　　　　　手表

香烟　　　　　　　　　　　　　　机上免税商品

化妆品　　　　　　　　　　　　　购物指南

首饰　　　　　　　　　　　　　　信用卡

酒类

【小组合作探讨完成下列任务】

写一写：分组讨论撰写"机上进行免税品销售，同时为乘客 Mike 介绍免税品——中国丝巾"的对话。

要　　求：(1) 符合为乘客提供免税品销售服务的流程；

(2) 要求情节设计符合逻辑，不少于 12 句；

（3）语法合乎要求，鼓励创新。

_____

_____

_____

_____

【角色扮演】 我要买点免税商品。（为乘客 Mike 介绍机上免税商品）

演一演： 分组角色扮演操练（实训场地）

要　求：（1）可使用自制的小道具进行展示，达到完成服务工作的目的；

（2）在展示过程中注意为乘客提供免税品销售服务的流程和礼貌服务；

（3）每组表演时间不超过5分钟；

（4）2～4人为小组代表在实训区进行展示；

（5）完成小组模拟情境互评打分表。

评价内容	专业能力 70分						职业素养 30分									总分	建议
组别	对话撰写无误，机上进行免税品销售流程清晰完整（30分）			表演中，成员发音清晰，语音语调准确，流程完整，表演流畅（40分）			对话表演中积极参与，情绪饱满，动作礼仪规范（15分）			对话表演有一定的创新性（5分）			对话表演中整体效果良好（10分）			100分	
得分	30	20	10	40	30	20	15	10	5	5	3	1	10	7	4		
一组																	
二组																	
三组																	
四组																	
五组																	

任务拓展和提升

【对话学习】Duty free shopping　免税品销售

FA：Sir, we're selling duty free items now. We have perfume, cigarette, cosmetics, accessories, alcohol, watch, chocolate and so on. This is our Duty Free Shopping Guide. You can find them on it. If there is anything you need, please let me know.

PAX：I want to buy some cigarettes.

FA：Sure. According to the entry requirements of America，each passenger can carry 400 cigarettes into America. So you can buy 2 cartons of cigarettes.

PAX：OK，2 cartons. I also want to buy something for my wife. Which one is suitable for lady?

FA：Perfume and cosmetics will be good for lady. Such as Christian Dior，Chanel，Estee Lauder and Clinique. All of these brands are popular around the world. Watches and accessories are also good.

PAX：OK. Maybe she likes this Miss Dior Cherie. I'd also like to buy some liquor. Three bottles of Royal Salute scotch whiskey，please. Oh，how much does it cost?

FA：$29 for each. But sorry，sir. Visitors to United States are allowed one liter at most.

PAX：Alright. Just one bottle. Can I have a discount?

FA：Sorry. All the items offered on board are sold at market prices. Your total price is $154.

PAX：Ok. How can I pay for them? Can I play it by cheque?

FA：I am afraid not. We only accept cash and credit card. For cash，we only accept US dollars，Euro and RMB. And the credit cards of VISA and Mastercard are also acceptable.

PAX：OK. I'll pay in cash by US dollars. Here is $200.

FA：I'm sorry，I don't have the change right now. Could I come back later after the Duty free sale? I will see if I can acquire change from other passengers'purchases.

PAX：That's fine.

FA：Thank you for your support and understanding.

Key words & expressions

item	n. 一件商品（或物品）	perfume	n. 香水
cigarette	n. 香烟	cosmetics	n. 化妆品
accessories	n. 首饰	alcohol	n. 酒
brand	n. 品牌	cheque	n. 支票
acquire	vt. 取得	purchase	v. 购买
support	vt. 支持	credit card	信用卡

【巩固练习】

1. 分角色朗读上面的对话。

2. 根据对话，回答下列问题。

1）Who are the two persons in the Dialogues?

2) What kind of duty free item does the flight offer?

_____

3) What duty free item did the passenger buy?

_____

4) How much are the duty free items?

_____

5) How did the passenger pay the item?

_____

3. 请将下列英文句子翻译成中文。

1) The perfume costs 41 dollars.

_____

2) The scarves are 72 dollars each.

_____

3) Forty-one plus 72 makes 113 dollars.

_____

4) A hundred dollars minus 85—that's 15 dollars change.

_____

5) That comes to 120 Euros.

_____

6) How will you be paying? By card or with cash?

_____

7) How would you like to pay?

_____

8) Here's your receipt，your card and your gifts.

_____

4. 听老师朗读，写出听到的单词或句子。

1) _____   2) _____
3) _____   4) _____
5) _____   6) _____
7) _____   8) _____
9) _____   10) _____

5. 请将下列免税品按照下面表格的要求分类。

whisky   a brooch   a USB key   face cream   cigars   perfume spray

cognac   aftershave   a soft toy   chocolates   earrings   lipstick

a pen   eau de toilette   vodka   a watch   a model aircraft   a scarf

champagne   mascara   a travel plug adaptor   a crystal pendant a travel razor   a bracelet cigarettes   headphones

Perfumes & jewelry	Electric and electronic items	Alcohol and tobacco	Cosmetics	Gifts

6. 朗读下面的广播稿，注意语音语调。

Ladies and Gentlemen：

Our duty-free shop will be open soon. We accept cash cards and major credit cards. For further information，please refer to the sales magazine in your seat pocket.

Thank you!

任务总结和知识点链接

1. 写出机上进行免税品销售基本服务流程。

2. 机上免税品销售结账途径。

3. 知识点链接。

机上免税品可进行销售的种类有香水、香烟、化妆品、首饰、酒类和手表等，现在多家航线和航空公司都推行了网上进行销售品销售的服务，乘客可以通过线上微信等多种渠道进行免税品的预购。

【课后练习和任务】

【课后练习】

1. 请将下列英文句子翻译成中文。

1）Could you recommend something as a gift for my wife?

2）Where was it made?

_____

3）All the items sold on board are tax free.

_____

4）How about this floral design? It's a conventional design and the colors are bright.

_____

5）I know that Hangzhou is famous for silk.

_____

6）That's just what I need.

_____

7）Keep the change.

_____

8）Can you show me some silk scarves?

_____

9）Here is your change.

_____

2. 请在横线上填上合适的英文单词或句子，使对话完整通顺。

CA：Would you like to _____ any duty-free items，sir?

PAX：Hmm，let me think. I'll have 200 Marlboro cigarettes and litre _____ of Black Label，please.

CA：Ok，sir. That will be 30 _____ . How would you like to _____ ?

PAX：Credit card. Do you _____ American Express?

CA：Of course，sir. Just a _____ ，please... I'm sorry，sir. But your card has

_____ .

【课后任务】

分小组，根据下面的背景编写对话，要求英文对话不少于 **12** 句，并将对语录制成视频以小组形式上传到学习通平台。

A man want to buy a souvenir to his wife in a plane，the attendant is helping the passenger to give some suggestion.

_____

_____

_____

_____

_____

任务描述

飞机平飞过程中，乘客 Anna 女士感觉不舒服，头晕，她觉得自己有些晕机，假如你是乘务员 Alice，请为这位特殊乘客进行服务。

任务准备

【任务相关知识准备】请熟记：特殊旅客的英文缩写。

1）VIP（Very Important People）

2）CIP（Commercially Important People）

3）UM（Unaccompanied Minor）

4）Special passengers（特殊旅客）：VIP，CIP，UM，senior passengers（年长乘客），expectant mothers（孕妇），passenger with disabilities and adult passengers with an infant or a small child（残疾人和携带婴幼儿的乘客）

5）minor 儿童，小孩

6）disabled passengers 残疾乘客

【小组查询资料并完成讨论】机上乘务员可为特殊乘客提供的服务有哪些？

任务实施

【梳理词汇和句型】 请熟记下列句型。

1. 安抚情绪	
I'm so/terribly/awfully/really/sorry.	
I'd like to _____ (道歉) for...	
Please forgive me for...	
Please _____ (接受) my sincere (deep) apology.	
So sorry have given you so much _____ (麻烦).	
2. 为什么?	
Due to	air _____ control (航空管制)
For	Mechanical reasons
Because	We add _____ (餐食) for some additional passengers.
3. 我们怎么做?	
I'm sorry we have to wait until the storm is over/the rain stops/the _____ (天气) improves.	
We're waiting for a few passengers to come on board/the _____ (飞机) ahead of us to take off/the aircraft to be loaded.	

【小组合作探讨完成下列任务】

写一写：分组讨论撰写"为生病乘客 Anna 提供帮助"的对话。

要 求：(1) 符合为特殊乘客（生病）服务的流程；

(2) 要求情节设计符合逻辑，不少于 12 句；

(3) 语法合乎要求，鼓励创新。

_____

_____

_____

_____

_____

_____

_____

【角色扮演】 我需要帮助。（为晕机乘客 Anna 提供帮助服务）

演一演：分组角色扮演操练（实训场地）。

要 求：(1) 可使用自制的小道具进行展示，达到完成服务工作的目的；

(2) 在展示过程中注意为乘客提供帮助服务的流程和礼貌服务；

(3) 每组表演时间不超过 5 分钟；

（4）2～4人为小组代表在实训区进行展示；

（5）完成小组模拟情境互评打分表。

评价内容	专业能力 70分						职业素养 30分									总分	建议
组别	对话撰写无误，为乘客提供帮助服务流程清晰完整（30分）			表演中，成员发音清晰，语音语调准确，流程完整，表演流畅（40分）			对话表演中积极参与，情绪饱满，动作礼仪规范（15分）			对话表演有一定的创新性（5分）			对话表演中整体效果良好（10分）			100分	
得分	30	20	10	40	30	20	15	10	5	5	3	1	10	7	4		
一组																	
二组																	
三组																	
四组																	
五组																	

任务拓展和提升

【对话】Dialogue A　Take care of ill passengers　照顾生病乘客

PAX：Excuse me, miss. I fell like vomiting.

FA：Don't worry. You may remove the armrest or adjust the seat back to lie down and relax. A good nap would be great for you. If you feel sick, you can use the airsickness bag in the seat pocket in front of you. And let me get you a cup of warm water.

PAX：I'm not feeling very well. Do you have airsick tablet?

FA：Yes, we do. But the tablet won't start to help so quickly. It works best 30minutes after usage. And for your health, we don't suggest you take any pills on board. But if you insist, we will get it for you, but you have to sign you name on the form (Letter of Consent of Use of Drug). Do you still need it?

PAX：Yes. I think I really need one.

FA：Sure, I will get it for you right now.

（5 minutes latter）

FA：Miss, here is the table and a cup of warm water. Please sign your name on the form.

PAX：Thank you. May I get a blanket? I want to take a nap.

FA：Sure. Here is your blanket and pillow.

PAX：It's really considerate of you.

Dialogue B   Passengers with Babies   带婴儿的乘客

FA：Excuse me，sir. This is a seat belt for your baby. She will feel more comfortable with it and it will be safer during an emergency. Let me connect it with your seat belt and you may adjust it as necessary.

PAX：Wow，you're so nice. By the way，could you please make a bottle of baby formula for my baby?

FA：Sure. How much hot water do you need?

PAX：200 mL will be fine.

FA：Sure! Just a moment.

(after a while)

FA：Here is the baby milk，sir. And if you need to change diapers for baby，the nappy changing tables are located in the rear lavatory of the left side.

PAX：Thank you. You've been very helpful.

Key words & expressions			
vomit	v. 呕吐	remove	v. 去除，脱掉
nap	v. 小睡打盹	airsick	n. 晕机
tablet	n. 药片	insist	v. 坚持
blanket	n. 毯子	considerate	adj. 体贴的
comfortable	adj. 舒服的	diaper	n. 尿布
lavatory	n. 卫生间	baby formula	婴幼儿配方奶粉

【巩固练习】

1. 分角色朗读上面的对话。

2. 根据对话 A，回答下列问题。

1）Who are the two persons in Dialogues A?

_____

2）What's wrong with the passenger in Dialogue A?

_____

3）How does the attendant help the passenger?

_____

4）What thing did the passenger sign in Dialogue A?

_____

3. 根据对话 B，回答下列问题。

1）Who are the two persons in Dialogue B?

_____

2）What kind of passenger in the plane in Dialogue B?

_____

3）How does the attendant help the passenger?

_____

4. 请将下列提出帮助的英文句子翻译成中文并熟读。

1）I'll check on our arrival time and get back to you.

_____

2）I'll ask if there is a doctor or nurse on board.

_____

3）I'll get you a blanket

_____

4）I'll get it for you.

_____

5）I'll show you how it works.

_____

5. 听老师朗读，写出听到的单词或句子。

1）_____ 2）_____
3）_____ 4）_____
5）_____ 6）_____
7）_____ 8）_____
9）_____ 10）_____

6. 朗读下面的广播稿，注意语音语调。

1）Ladies and Gentlemen：

May I have your attention, please? There is a sick passenger on board. If there is a doctor or nurse among you please contact us by pressing the call button immediately.

Thank you!

2）Ladies and Gentlemen：

Due to turbulence weather, please return to your seat and fasten your seat belt. Parents please lift your babies out of the bassinets. Use of the lavatories will be suspended during the turbulence.

Thank you!

3）Ladies and Gentlemen：

Due to turbulent weather，we'll suspend in-flight service（lunch/dinner... service）until the weather is clear.

We're sorry for the inconvenience caused.

Thank you！

任务总结和知识点链接

1. 为特殊乘客提供的服务有哪些？

2. 知识点链接：一些特殊旅客的相关服务。

无成人陪伴儿童

东航有一项服务，叫作"小小旅行家"，是专为5～12周岁、无成人陪伴单独乘机的儿童提供的。

办理的手续并不繁杂：只需要提前一周到东航售票处办理"无成人陪伴儿童"乘机申请，待确认了对方接机人后就可以购买机票。到达机场前，家长可拨打东航的"温馨预约"电话进行预约，并告知小旅客到达时间，东航的服务员会一路带领小旅客办理各类手续。由于UM旅客的年龄较小，每一位小旅客都有"温馨姐姐"提供服务。在整个登机过程中，"温馨姐姐"将始终陪同小旅客下棋、看儿童读物、讲故事等，为"小小旅行家"们提供良好的服务。

登机时，服务员会把小旅客交给乘务员，乘务员会在飞机上对小旅客进行精心照顾；下机时，已收到电报的到达站服务员会准时来登机口迎接，帮助办理相应的手续，直到把小旅客完好地交给家人。

"小小旅行家"们不仅一路有东航的叔叔阿姨精心呵护，还会收到一张奖状作为这趟"勇敢旅行"的永久纪念。孩子需要独立，也需要关心。东航"小小旅行家"，大人放心，小孩开心。

轮椅服务

"轮椅走天涯"是东航专为病残或行走不便的旅客量身定做的服务项目。如果您的家人需要东航提供轮椅服务，可在订座时提出，您的需求信息会从售票处转至机场，也可拨打"温馨预约"电话和我们联系，东航的工作人员会备好轮椅准时在柜台恭候，并将一路协助办理乘机手续、安检、边防等手续，工作人员还会与飞机上的乘务员交接，让旅客在飞机上受到乘务远的照顾。到了目的地，服务员会提前备好轮椅来迎接。

怀抱婴儿专座

在东航飞机上，有一类特殊的座位，是专门为怀抱婴儿的旅客设计的。这些座位往往比一般座位间距要宽，座位前隔离板上的两个特别铆钉可以用来放置婴儿摇篮。当然，最重要的是，只要乘客在购票时提供信息，售票员将婴儿信息输入电脑，东航的工作人员会提前给乘客预留这种婴儿座位。

老人服务

年老旅客行动不方便，单独乘机可能有一定的困难。东航特别为年老旅客安排专人服务，一路陪同直至登机，其间更是为老人解决相应的问题。

【课后练习和任务】

分小组，根据下面的背景编写对话，要求英文对话不少于 12 句，并将对语录制成视频以小组形式上传到学习通平台。

A man feel stomachache in a plane, the attendant is offering help to the passenger.

_____

_____

_____

_____

_____

_____

_____

情景二　Cabin Service 客舱服务

学习活动三　飞机着陆服务（Landing Service）

　　飞机已经到达目的地机场，假如你是空中乘务员 Alice，请在舱门口处欢送乘客 Johnson 下机，并帮助其拿好手提行李。

任务准备

【小组查询资料并完成讨论】 利用电脑、手机等媒介，查询资料并讨论：

1）飞机即将到达目的机场，正在下落，此时乘务员应该提醒乘客做什么？

2）飞机已经降落到地面，但仍在滑行阶段，此时乘务员应该提醒乘客做什么？

3）飞机已经平稳着陆，乘务员准备在舱门处送别乘客时应该注意什么？

任务实施

乘务员在飞机着陆后的服务程序：

【小组合作探讨完成下列任务】

写一写：分组讨论撰写"在舱门处欢送乘客 Johnson 下机"的对话。

要　求：（1）符合为欢送乘客下机的流程；

　　　　　（2）要求情节设计符合逻辑，不少于 12 句；

　　　　　（3）语法合乎要求，鼓励创新

_____

_____

_____

_____

_____

_____

_____

_____

_____

【角色扮演】到成都了！（在舱门口处欢送乘客 Johnson 下机）

演一演：分组角色扮演操练（实训场地）。

要　求：（1）可使用自制的小道具进行展示，达到完成服务工作的目的；

　　　　　（2）在展示过程中注意为欢送乘客下机的流程和礼貌服务；

　　　　　（3）每组表演时间不超过 5 分钟；

　　　　　（4）2～4 人为小组代表在实训区进行展示；

　　　　　（5）完成小组模拟情境互评打分表。

评价内容	专业能力 70 分						职业素养 30 分									总分	建议
组别	对话撰写无误，欢送乘客下机流程清晰、完整（30 分）			表演中，成员发音清晰，语音语调准确，流程完整，表演流畅（40 分）			对话表演中积极参与，情绪饱满，动作礼仪规范（15 分）			对话表演有一定的创新性（5 分）			对话表演中整体效果良好（10 分）			100 分	
得分	30	20	10	40	30	20	15	10	5	5	3	1	10	7	4		
一组																	
二组																	
三组																	
四组																	
五组																	

课后拓展和提升

1. 朗读下面的广播稿，根据实际情况在横线上填写合适的单词。

Ladies and Gentlemen：

Our plane has landed at _____ airport. The local time is _____ . The temperature outside is _____ degrees Celsius (_____ degrees Fahrenheit). The plane is taxiing. For your safety, please stay in your seat for the time being. When the aircraft stops completely and the Fasten Seat Belt sign is turned off, Please detach the seat belt, take all your carry-on items and disembark (please detach the seat belt and take all your carry-on items and passport to complete the entry formalities in the terminal). Please use caution when retrieving items from the overhead compartment. Your checked baggage may be claimed in the baggage claim area. The transit passengers please go to the connection flight counter in the waiting hall to complete the procedures. Welcome to _____ (city), Thank you for selecting × × airline for your travel today and we look forward to serving you again. Wish you a pleasant day. Thank you!

2.【对话】Dialogue A　Seeing Passengers Off at the Cabin Door 在舱门处送别乘客

Pur：(Announcement) Ladies and gentlemen, we have just landed at Shanghai Hongqiao Airport. Please do not unfasten your seat belts until the plane comes to a complete standstill. Please, make sure you collect all your belongings before you disembark. Your checked baggage may be claimed in the terminal building.

A_1：Goodbye, Mr. Wang, have a nice day!

P_1：Thanks.

A_2：See you, madam, and I hope you have a good time here!

P_2：I will. Thank you.

A_3：Hey, Tommy. Did you have a nice trip?

P_3：Yes, I think so. Thank you for the toys and games.

A_3：You're welcome. Give my regards to your family.

A_4：Bye, sir. Thank you for flying with A Airline.

P_4：Oh, thank you, too. I appreciate your service very much. I'll write a thank-you letter to your airline when I get home.

A_4：Many thanks indeed.

Dialogue B   Introducing a City to a Foreign Passenger
向国外乘客介绍某一城市

P：Excuse me，Miss. Could you tell me what the time is in Chengdu now?

A：It's half past ten in the morning. Chengdu time is thirteen hours ahead of New York.

P：I see. Thank you. This is my first visit to China. I'll be in Chengdu in a few hours. I'm very excited.

A：Why are you so excited?

P：As is well known to the world，China has rich historical and cultural heritage and something mysterious to the foreigners.

A：What's more，so many historical and scenic spots attract tourists from all parts of the world.

P：Yes，you are quite right. I hear there are lots of interesting places in Chengdu. Could you tell me some of them?

A：Yes. Chengdu is one of the oldest cities in China. It has Dujiangyan，Qingchengshan，Sanxingdui and so on.

P：That must be very interesting. And I want to see Dujiangyan with my own eyes. What do you think about it?

A：That's right. And if you have more time you could also stroll along the Chengdu Hutong. That'll be also very interesting.

P：Really?

A：Yeah，I hope you have a nice trip in Chengdu.

P：Thank you very much for the information.

Key words & expressions

land	v. 着陆，到达	complete	adj. 完全的
standstill	n. 停止	collect	v. 收集
belonging	n. 物品，所属物	appreciate	v. 感谢
regard	v. 问候	indeed	adj. 真正地
visit	v. 参观	excited	adv. 激动地
historical	adj. 历史的	mysterious	adv. 神秘的
foreigner	n. 外国人	scenic	n. 风景
attract	v. 吸引	tourists	n. 游客
interesting	adj. 有趣的	trip	n. 旅行
stroll	v. 漫步，闲逛	information	n. 信息
see off	给……送行	ahead of	提前
cultural heritage	文化遗产		

【巩固练习】

1. 分角色朗读上面的对话。

2. 根据对话，回答下列问题。

1) Who are the two persons in the dialogue A?

_____

2) Where does the passenger want to go in dialogue B ?

_____

3) How did the passenger 4 want to do in Dialogue A?

_____

4) How did the attendant help the passenger in dialogue B?

_____

3. 听老师朗读，写下所听到的单词或句子。

1) _____   2) _____

3) _____   4) _____

5) _____   6) _____

7) _____   8) _____

9) _____   10) _____

4. 按要求完成下列习题。

1) 在横线上填写合适的单词、短语或句子，使对话通顺。

A₁: Goodbye, Mr. Wang, _____ a nice day!

P₁: Thanks.

A₂: See you, madam, and I hope you have a _____ _____ here!

P₂: I will. Thank you.

A₃: Hey, Tommy. Did you have a nice _____?

P₃: Yes, I think so. _____ _____ for the toys and games.

A₃: You're welcome. Give my _____ to your family.

2) 将下列句子翻译成英文。

(1) 本架飞机已经降落在上海虹桥机场。

_____

(2) 您交运的行李请到候机室领取。

_____

(3) 祝您愉快。

_____

(4) 感谢您乘坐 A 航空公司。

_____

(5) 等飞机完全停稳后，请您再解开安全带。

_____

5. 在横线上填写合适的单词、短语或句子，使对话通顺。

P：Excuse me, Miss. Could you _____ me what the time is in Chengdu now?

A：It's half past ten in the morning. Chengdu time is thirteen hours _____ _____ New York.

P：I see. Thank you. This is my first _____ to China. I'll be in Chengdu in a few hours. I'm very _____ .

A：Why are you so excited?

P：As is well known to the _____ , China has rich _____ and cultural heritage and something mysterious to the _____ .

A：What's more, so many historical and scenic spots _____ tourists from all parts of the world.

6. 将下列中文句子翻译成英文或将英文句子翻译成中文。

1）这是我第一次来中国。

_____

2）众所周知，外国人对于有着历史文化遗产丰富的中国，总有一种神秘感。

_____

3）我听说在成都有许多的名胜古迹。

_____

4）如果您的时间充足，建议您可以逛逛成都的胡同。

_____

5）Chengdu time is thirteen hours ahead of New York.

_____

6）I want to see Dujiangyan with my own eyes.

_____

7. 请大声朗读下面的广播稿。

Ladies and Gentlemen：

The plane has stopped completely；please disembark from the front (middle，rear) entry door. Thank you!

任务总结和知识点链接

飞机着陆后，乘务员的服务流程：

1）广播欢送词：航班延误应再次向旅客表示歉意；飞机未停稳前，提醒旅客不要开启手机电源，劝阻旅客不要起立或者打开行李架。

2）灯光调至最亮，调节时应注意从暗到亮。

3）乘务长统一指挥操作舱门分离器（滑梯）解除预位。

4）播放音乐。

5）打开舱门：①检查客梯/廊桥到位，接到地面工作人员"可以开门"信号后打开舱门；②开门前，两人再次确认舱门分离器已经解除预位后再开门。

6）欢送旅客：着装整齐、礼貌道别；安排 VIP、头等舱、公务舱旅客先下飞机，协助特殊旅客下飞机。

7）检查客舱：逐个检查行李架和旅客座位。如发现旅客遗留物品，立即报告主任/乘务长，通知地面有关人员。

课后练习和任务

【课后练习】
将下列英文句子翻译成中文或将英文句子翻译成英文。

1）Did you have a nice trip?

_____

2）Give my regards to your family.

_____

3）I appreciate your service very much. I'll write a thank-you letter to your airline when I get home.

_____

4）What's more, so many historical and scenic spots attract tourists from all parts of the world.

_____

5）I hope you have a nice trip in Chengdu.

_____

6）Thank you very much for the information.

_____

7）成都是中国最古老的城市之一，在成都有都江堰、青城山和三星堆遗址等。

_____

8）你觉得怎么样呢？

_____

9）确实很感谢！

_____

10）希望您在这里过得愉快！

_____

11）等飞机完全停稳后，请您再解开安全带。

_____

12）请整理好您的手提物品，准备下飞机。

_____

【课后任务】

分小组，根据下面的背景编写对话，要求英文对话不少于 12 句，并将对语录制成视频以小组形式上传到学习通平台。

You want to land at A airport，please ask the attendant help you to take care of your kid，and you are ready to take your belongings.

_____

_____

_____

_____

_____

_____

_____

_____